THE FOUR POWERS

THE
FOUR POWERS

JASON M. HANANIA

San Francisco

FIRST EDITION, JULY 2023

Library of Congress Control Number: 2023911419

ISBN (hardcover): 978-1-7321197-6-5
ISBN (paperback): 978-1-7321197-9-6
ISBN (epub): 978-1-7321197-2-7
ISBN (mobi): 978-1-7321197-5-8
ISBN (audiobook): 978-1-7321197-8-9

Published in the United States by www.4powers.com.

Please contact us using media@4powers.com.

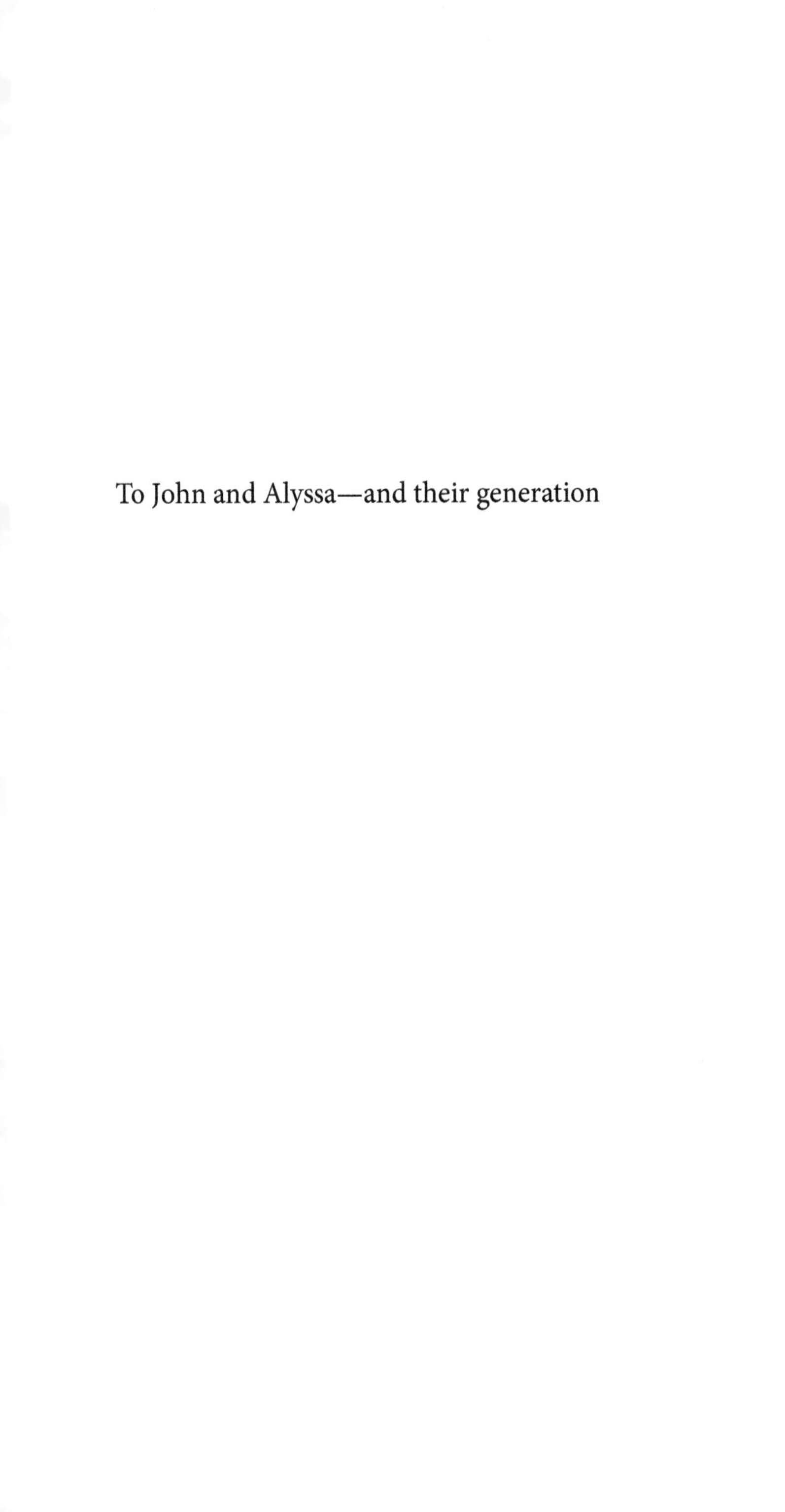

To John and Alyssa—and their generation

TABLE OF CONTENTS

EXPANDED TABLE OF CONTENTS

INTRODUCTION

An Equation for Power

In 2016 I set out to write the book *Architecture of a Technodemocracy*. It documents a system for ending political party systems using technology and democracy. Before authoring that book, I searched far and wide for an institutional definition of democracy but found none.

If you asked ten social scientists for a simple definition of democracy, you would get ten different responses. As an engineer, I found this frustrating. I wanted a more technical understanding of democracy, a mathematical framework to build on. The same way physical science provides equations for motion (such as force equals mass times acceleration, or $F = ma$), I wanted a social science equation for democracy.

After struggling with this problem for months, I concluded that democracy is the decentralization of governing power. Subsequently, I needed to define governing power (G_{Power}). The end result was a simple equation:

$$\mathbf{G_{Power} = C + O + D + A}$$

C = communication power
O = option power
D = decision power
A = accountability power

The four powers, CODA, apply to any group, be it planet, nation, state, town, church, business, team, or family. They are like the four components of a sound system—volume, bass, fade, and treble—that can be adjusted individually to attain the preferred balance. As shown below, the four powers can shift together or individually from being centralized in one group member to being fully decentralized.

Communications:	0%---<I> ----------100%
Options:	0%----<I> ---------100%
Decisions:	0%--<I> -----------100%
Accountability:	0%-<I> ------------100%

As with a sound system, people tend to disagree over what constitutes the ideal settings.

Defining the Four Powers

A government is not a white marble building filled with politicians, administrators, businessmen, religious leaders, or other group leaders. It is a collection of social mechanisms (i.e., people processes). The four powers drive those mechanisms. For any group, effective governance requires that members engage in the processes of *communicating*, weighing *options*, making *decisions*, and *accounting* for those decisions. The four powers equation allows us to apply calculus to governance.

At a high level, the minimum and the maximum show us there are two clear ends on the government spectrum: democratic and nondemocratic. In a nondemocratic government, these four powers are typically *centralized* in one group member or less than 1% of group members (the 1%). In a democratic government, the four powers are *decentralized* to up to 100% of group members (the 100%). The concepts of centralizing and decentralizing power are fundamental to understanding government.

Through the decentralization of power, democracy offers *equality*. Every group member is vested with equal power.

Each of the four powers is briefly discussed below.

1. Communication Power

Communication power is a function of *connectivity* and *opportunity*. Group members have greater opportunities when they are connected to influential, intelligent, attractive, creative, athletic, or otherwise wealthy group members. Birth parents are a human group member's first two connections. Communication power is the focus of Part I of this book.

2. Option Power

Option power is a function of *resources* and *wealth*. Wealthy group members have more options because they have more resources. Impoverished group members have fewer options because they have fewer resources. The adage "money is power" is too narrow. Option power certainly includes money and other financial resources, but it also includes human resources (such as labor, ideas, and talent), natural resources (land, water, gold, and so on), and information resources (educational, experiential, evidentiary, and the like). Option power is the focus of Part II of this book.

3. Decision Power

Decision power is a function of *agreements* and *freedom*. Freedom is the ability to form your own agreements rather than have another group member—such as a parent, spouse, priest, politician, campaign donor, corporate executive, employer, or king—influence or make decisions for you. If another group member controls your options, they control your decisions. Decision power is the focus of Part III of this book.

4. Accountability Power

Accountability power is a function of *evidence* and *responsibility*. Evidence enables accountability over decisions and is collected through overt methods (such as open investigations) and covert efforts (such as spying). It is heavily dependent on documentation (audio, video, testimony, contracts, prints, checks, receipts, spreadsheets, and so on). Group members have a responsibility to communicate accountability information, or to whistle blow, when a policy, guideline, rule, protocol, code, standard, contract, statute, law, or other agreement is compromised. Accountability power is the focus of Part IV of this book.

A Framework for Understanding Power

Historically, groups tend to deal with social problems, such as those arising from race, gender, religion, wealth, immigration, climate, education, or healthcare issues, by using problem-tight compartments. The four powers equation allows us to see that centralized power, or *power inequality*, is the overriding cause of most social problems. This suggests that addressing power inequality as a whole, rather than managing social problems as individual and compartmentalized situations, offers the most efficient path to democracy, equality, and reduced social risk.

The four powers framework is substantively consistent and includes the following examples:

C: connections (churches, families, schools, employers, other networks, etc.)
O: resources (spiritual, human, natural, financial, informational, metaphysical, etc.)
D: agreements (legislation, contracts, votes, priorities, purchases, ratings, etc.)
A: evidence (self-observation, testimony, videos, documents, ledgers, etc.)

The four powers framework is also procedurally consistent. Consider the mechanisms and examples below:

C: media (journalists, clergy, producers, pundits, salespersons, etc.)
O: economics (bankers, entrepreneurs, engineers, artists, entertainers, etc.)
D: politics (parents, teachers, priests, government officials, CEOs, etc.)
A: security (police, attorneys, accountants, paramedics, soldiers, etc.)

The four powers framework ultimately illuminates the primary rights required within any democratic government system:

C: the right to communicate (religion, speech, press, assembly, protest, etc.)
O: the right to options (food, water, shelter, healthcare, energy, information, etc.)
D: the right to decide (living, dying, marrying, reproducing, immigrating, voting, etc.)
A: the right to accountability (whistleblowing, auditing, investigating, impeaching, etc.)

Any individual capable of successfully asserting each of the four powers is presumably a viable group member.

A Framework for Understanding Leadership

Any group member, at any time, is capable of leading the group. The "alpha male" is a misnomer. The alpha in any group is the most powerful group member *specific to circumstances*. It's situational. The most powerful group member in any situation is the one offering the best option for the group, regardless of gender, race, wealth, or otherwise, under the given circumstances. Anybody can be an alpha.

Strong leaders serve the group by:

C: communicating well and genuinely connecting with as many group members as possible
O: providing financing, labor, ideas, motivation, or other resources to the group
D: making decisions that agree with majority interests rather than only considering self-interests
A: holding all group members accountable, including themselves, based on evidence

For example, a king can be democratic if he implements the four powers in the following ways:

C: he communicates with all of his people,
O: he does not hoard life-essential resources,
D: his decisions reflect majority interests, and
A: he is accountable to the laws of the kingdom.

On the other hand, a republic can be nondemocratic if any of the following situations arise:

C: Elected officials cease communicating with constituents following elections.
O: Minority interests, such as corporations, hoard resources.

D: Minority interests, such as campaign donors, influence decision-making.

A: Elected officials are shielded, such as by political party, and effectively operate above the law.

Being crowned king, elected to office, appointed CEO of a corporation, ordained as a priest, hired as a head coach, or placed at the head of any other group does not necessarily make someone a leader. It simply means that person is an empowered representative of the group. These representatives abuse their power when their decision-making prioritizes self-interests (financial, sexual, egoic, and the like) over group interests (social, economic, environmental, and so on). As such, anticipating abuses of power becomes an exercise in spotting *conflicts of interest.*

Overview

Using historical examples, *The Four Powers* provides a concise and fast-paced look at power inequality, conflicts of interest, and abuses of power. Part I focuses on communication power from 1940 to 1960, Part II focuses on option power from 1960 to 1980, Part III focuses on decision power from 1980 to 2000, and Part IV focuses on accountability power from 2000 to 2020. Readers are encouraged to research the allegations and deductions herein and to keep an open mind. Prevalent concepts include the following:

C: warrantless surveillance, mass media, disinformation, spin, distractions, lies, false flags, the cherry-picking of facts, censorship, the controlling of narratives, and other propaganda techniques

O: central banking, hoarding, wealth inheritance, tax evasion, price fixing, corporate monopolies, product duopolies, product assassination, political party duopolies, political assassination, intimidation, and other option control techniques

D: enslaving, imprisoning, killing, or committing other acts of violence; creating financial incentives or other conflicts of interest; offering or receiving campaign donations, kickbacks, bribes, or other quid pro quo; participating in gerrymandering or election rigging; vetoing or criminalizing under pretext; pardoning coconspirators; and taking part in other electoral, executive, legislative, or judicial decision-rigging techniques

A: information control (through classification, top secrets, trade secrets, patents, confidentiality, privileges, immunities, and other laws), covert agencies, shell corporations, identity fraud, self-regulation, retaliation (voter, whistleblower, or juror), cover-ups, and other anti-accountability techniques

By understanding the four powers, readers will be better equipped to: (1) equalize power through technodemocratic social mechanisms, (2) anticipate abuses of power by spotting conflicts of interest, and (3) provide for national and global security through reduced social risk.

Democracy on Earth is still very much an experiment. Like cryptocurrency, technodemocratic social mechanisms are now being implemented without the help of politicians, corporations, bankers, churches, or any other middlemen. The possibility exists for blockchain-based national and global security. One software developer can change the world.

—JASON M. HANANIA

PART I

COMMUNICATION POWER

CHAPTER 1

Propaganda

The Hero

On July 13, 1894, in a dark, smoke-filled tenement on a Bosnian farm, a baby boy was born onto a cold stone floor. The boy's mother was an impoverished Serb and was denied access to Bosnian hospitals. Unable to afford basic healthcare resources, the boy's grandmother had to sever the umbilical cord with her teeth. Hearing the cries of a newborn, the entryway became a chokepoint for onlookers.[1]

The boy's mother, Marija Princip, wanted to name him Spiro, after her late brother. But a priest stepped in—it was not a woman's place to decide the name. According to the priest, the boy needed a good Christian name. He was born on the day of Saint Gabriel, so his name would be Gavrilo. Gavrilo Princip was the second of nine children, six of whom would die in infancy from the likes of dysentery, fever, and pneumonia.[2, 3]

The Princips lived in a region oppressed by the Austro-Hungarian Empire. Under the feudal system, the Princips were not slaves. They were not owned by another person. Instead, they were serfs, fixtures of the land, like mules or tractors. When land resources transferred between lords, the serfs transferred with it. By law, leaving your fixture was punishable by death.[4]

Gavrilo Princip was raised on stories of freedom and revolution. As a teenager, he and his peers were forbidden from communicating government accountability information, so they decided to form secret groups and seek out other revolutionaries. Gavrilo Princip regularly asserted his communication power and was eventually expelled from school over a peaceful protest.[5] Soon after, he shot and killed one of his oppressors.[6]

A leading Serbian cleric, Metro Amfilohije, declared that Gavrilo Princip was "defending his freedom and his people."[7] The way Serbian Chairman Milorad Dodik explained it, Gavrilo Princip was ready "to give his life for the freedom of his own people."[8] In a speech, Serbian President Tomislav Nikolic called Gavrilo Princip "a hero, a symbol of the idea of freedom."[9]

Gavrilo Princip died in a prison cell at age twenty-three. A century later, a bronze statue of him was erected in Belgrade. The larger-than-life tribute serves to inspire future revolutionaries and freedom fighters.

The Terrorist

On June 28, 1914, the archduke of Austria, Franz Ferdinand, was visiting Austro-Hungarian–occupied

Bosnia. Archduke Ferdinand was the nephew of the Austro-Hungarian emperor. By no achievement of his own, Ferdinand was next in line to the throne.

Bosnia was rich in natural resources, and the Austro-Hungarian Empire needed those resources to advance its conquest of Europe. Like the emperor, the archduke was born into great wealth—massive amounts of land, gold, and other resources. He also had the loyalty of the empire's army—well-paid soldiers who could violently enforce the laws that made it all possible.[10]

In resistance to the forced occupation of his country and the archduke's visit, a Bosnian insurgent tossed a bomb into a motorcade of Austro-Hungarian politicians accompanying the archduke. The injured victims were granted access to a Bosnian hospital, and the archduke decided to pay them a visit.

Archduke Ferdinand was chauffeured to the hospital in a Gräf & Stift Double Phaeton convertible. Due to a miscommunication, the Archduke's driver stopped on a side street. Standing on that street was a nineteen-year-old member of a secret group called Young Bosnia.[11]

Young Bosnia wanted the Austro-Hungarian Empire out of Bosnia and had been branded a terrorist organization by authorities. Recognizing who was in the convertible, the nineteen-year-old approached the vehicle, pulled a handgun, and shot Archduke Ferdinand in the neck. Onlookers cried out in terror as the archduke choked to death on his own blood.[12] The shooter was Gavrilo Princip.[13]

Igniting World War I

Propaganda is any idea communicated with the intent to influence the decision-making of others. If a decision is like a stick of dynamite, then propaganda is like a fuse. The decision in this case was war. The death of Archduke Ferdinand by the hand of young Gavrilo Princip ignited World War I. Princip's actions communicated an idea. That idea varied drastically depending on who propagated it.

At the time, communication power was centralized in newspaper and radio. There was no television or Internet. Newspaper and radio offered a mass media for propagating one-sided narratives. Like sculptors, group members wielding this type of communication power could craft ideas that influenced the decisions of the 99%. They could control the narrative. In the case of World War I, two competing narratives polarized the world.[14]

Austro-Hungarian propaganda spun Princip as a terrorist, an insurgent, an assassin, and a murderer. The assassination was an excuse for the Austro-Hungarian and German 1% to declare a war on terrorism. At the same time, Bosnian propaganda spun Princip as a hero, defender, freedom fighter, and liberator. The Russian, Belgian, French, and British 1% decided to defend Bosnia.[15, 16]

In each of those countries, a majority of the 99% took the bait. Using propaganda and centralized media, each country's decision-makers quickly persuaded their respective 99% that war was that country's best option. Mind you, these decision-makers, representing less than 1% of the population, would never set foot in combat. They were businessmen, and the war was just another transaction. They had personal financial interests influencing their decision-making.

On a steady diet of propaganda, members of the 99% in each nation were taken to slaughter. Like livestock, young men and women were resources, tools for the 1%.[17]

CHAPTER 2

False Flags

Igniting World War II: Japan's False Flag

On September 18, 1931, Japanese army lieutenant Suemori Komoto was secretly ordered to covertly bomb a section of railroad tracks *in his own country*. The bombing was a false flag, an attempt by the Japanese 1% to fool the Japanese 99% into believing that China was invading Japan. It worked. Centralized Japanese media propagated the idea that China had attacked an economically vital section of Japanese railroad.[18, 19, 20]

Japanese army officials had covertly orchestrated the false flag as a pretext to invade China. Chinese Manchuria was rich in untapped oil, iron, gold, timber, and other natural resources. Japanese army officials had centralized access to communicators (loyal media officials), human resources (loyal Japanese soldiers), and decision-makers (loyal politicians). They simply needed to maintain centralized accountability power by keeping the operation a secret from the public.[21, 22]

They succeeded. As a result of the false flag, the Japanese 99% decided that China was an imminent threat and thrust support behind Japan's decision-makers. Within hours of the so-called attack, Japan's decision-makers invaded Chinese-Manchuria under the guise of self-defense. China was caught flat-footed and had no way to communicate with and persuade the Japanese 99% that the Japanese 1% was lying to them.[23]

In Japan, lying was of course considered dishonorable. However, Japanese soldiers and government officials, who were aware the bombing was a false flag, had too much to lose by whistleblowing. They would be branded traitors, discredited as liars, and generally humiliated. Their careers would be over, and they would lose any means of supporting their spouses, children, and other family members. Ultimately, they would face imprisonment or even death. They had conflicts of interest. Their personal interests (protecting themselves and their families) conflicted with public interests (protecting other families). There were no democratic social mechanisms to prevent retaliation and protect whistleblowers.[24]

The undemocratic social mechanisms that helped make the Manchuria false flag successful did not arise overnight. In the decades leading up to the world wars, communication power (media) and accountability power (law enforcement) were deliberately centralized in the Japanese 1%. Utilizing spin words like "safety," "order," and "security," the Japanese 1% legislated three laws that would ensure its hold on power:

the "safety preservation" law of 1894, the "public order and police" law of 1900, and the "public security preservation" law of 1925.

Communicating government accountability information became a crime. The Kenpeitai, a secret police force, arrested roughly 50,000 people.[25] The penalties for whistleblowing ranged from ten years' imprisonment to death. Japan's false flag ultimately snowballed into the Pacific theater of World War II. In that region, approximately 20 million people were slaughtered.[26, 27]

Igniting World War II: Germany's False Flag

On August 31, 1939, German operatives wearing Polish uniforms staged a phony invasion of Gleiwitz, Germany. The operation was a false flag, an attempt by the German 1% to fool the German 99% into believing that Poland was invading Germany. It worked. A war prisoner was secretly dressed in a Polish soldier's uniform and murdered by German operatives. Photos of the dead body were then propagated through German media as evidence Poland was invading Germany.[28, 29]

The German 1% also propagated the idea that Polish troops had seized a German radio tower. Using the radio tower, a message sympathetic to Poland was communicated by German operatives, in Polish, to the German 99%. At the time, the German 1% had centralized control over German media. From the perspective of the German 99%, their country was under attack.[30, 31]

The German 1% had fooled the German 99% into deciding they needed to defend themselves. At the same time, Poland was caught flat-footed and had no way to communicate with and persuade the German 99% that the German 1% was lying. Within twenty-four hours, Germany invaded Poland and the European theater of World War II was underway. In that region, over 40 million people were slaughtered.[32, 33]

CHAPTER 3

Censorship

1941: How the 1% Drags the People Along

The German 1% was attempting to transform Germany into a global empire. In its infancy, the Nazi political party understood the need to centralize communication power by consolidating German media outlets. As a result, the Ministry of Public Enlightenment and Propaganda was created.

The minister of propaganda, Joseph Goebbels, then censored or banned any remaining independent media that was competing with Nazi Party–controlled media. This optimally centralized communication power in the German 1%. Goebbels would later confess, "Even during the time when we were in the opposition, we succeeded in rescuing the concept of propaganda from disfavor or contempt. Since then, we have transformed it into a truly creative art. It was our sharpest weapon in conquering the state."[34]

Nazi Party officials were overthrowing their own government. While Goebbels was turning the

German media into a mockingbird, Nazi Party official Hermann Göring was centralizing accountability power. Göring was building a covert agency called the Gestapo. The Gestapo would only be accountable to the Nazi Party officials who were running it. The German people would have no accountability power over the Gestapo. Göring would later confess how the German 1% dragged along the 99%:

> "Why, of course, the people don't want war. Why would some poor slob on a farm want to risk his life in a war when the best that he can get out of it is to come back to his farm in one piece? Naturally, the common people don't want war—neither in Russia, nor in England, nor in America, nor for that matter in Germany. That is understood. But, after all, it is the leaders of the country who determine the policy and it is always a simple matter to drag the people along, whether it is a democracy, or a fascist dictatorship, or a parliament, or a communist dictatorship.... Voice or no voice, the people can always be brought to the bidding of the leaders. That is easy. All you have to do is tell them they are being attacked and denounce the pacifists for lack of patriotism and exposing the country to danger. It works the same in any country."[35]

To that end, the Nazi Party utilized new technologies to drag the German people along, including Enigma machines and Volksempfängers. Enigma machines were cipher devices that centralized communication power by encrypting Nazi communications. Volksempfängers were cheap radios that could only receive broadcasts from the Nazi-controlled German State Radio. They did not receive foreign broadcasts. Volksempfängers connected the Nazi Party to every household and effectively censored enemy narratives. The Nazi Party had the option, at any moment, to override local music channels and propagate a Nazi Party–controlled narrative. The German 99% believed what they were hearing was the truth.[36]

1942: Centralizing Communication Power

The Nazi Party's minister of war, Albert Speer, later confessed,

> "Hitler's dictatorship differed in one fundamental point from all its predecessors in history. His was the first dictatorship in the present period of modern technical development, a dictatorship which made the complete use of all technical means for domination of its own country. Through technical devices like the radio and loudspeaker, 80 million people were deprived of independent thought."[37]

Communications technology has since evolved from the likes of local Nazi radio and loudspeakers to global media corporations. The same way connectivity enables abuse of communication power through censorship and the transmission of information (propaganda), it also enables abuse through the receipt of information (surveillance). Below are some examples.

C: AT&T, Verizon, China Telecom, China Unicom, MTC, and the like have centralized communication power because they control the Internet. At all times, these entities have a back door to global communications platforms and can secretly survey all communications.

O: Companies such as Amazon, Facebook, Twitter, Instagram, Douyin (TikTok), PayPal, and VKontakte have centralized communication power because they provide the most popular platforms and decide what gets propagated. Propagation power can be purchased. Through censorship, these entities can also block both products and government accountability information. In addition, they can secretly barter their users' personal information for revenue and tax breaks.

D: Fox News, CNN, Al Jazeera, the Guardian, the New York Times, Time magazine, CGTN, ITAR-TASS, RIA Novosti, and so on have centralized communication power because they provide the most popular sources for contemporaneous accountability information ("news"). These entities can influence group decisions by propagating facts that fit a particular narrative.

A: Apple, Google, Samsung, Huawei, Xiaomi, Yota, and others have centralized communication power because they manufacture virtually every communication device. These entities can perpetually account for a user's GPS location as well as almost every communication transmitted or received. This occurs using surveillance technology installed during or after the manufacturing process, such as through "software updates."

While communication power has been gradually decentralized from the likes of kings and dictators, it still remains heavily centralized in less than 1% of Earth's population.

CHAPTER 4

Surveillance

1943: Project Rubicon

Surveillance, without consent or warrant, is one way to abuse communication power. For decades under project Rubicon, the U.S. National Security Agency (NSA) and Central Intelligence Agency (CIA) used a fraudulent Swiss corporation called Crypto AG to manufacture and sell encryption machines to private buyers. The machines allowed for encrypted international communications and were popular among foreign leaders. What those foreign leaders did not realize was that NSA and CIA officials could remotely decrypt and read all the communications sent by any Crypto AG machine.[38, 39]

As revealed by whistleblowers like Edward Snowden and Mark Klein, many international corporations have a history of cooperating with secret government agencies like the NSA, CIA, KGB (Committee for State Security, Russia), FSB (Federal Security Service, Russia), SVR (Foreign Intelligence Service,

Russia), Gestapo (Germany), BND (Federal Intelligence Service, Germany), MI6 (Secret Intelligence Service, Britain, also known as SIS), RAW (Research and Analysis Wing, India), MSS (Ministry of State Security, China), and Mossad (Israel).

For any group, taking control of communications platforms is step one in the process of centralizing all four powers of government. To be truly democratic, communications platforms must be owned and controlled by 100% of their users, not a centralized group, such as a church, political party, or corporation.

Operation Hummingbird

Several years before the Gleiwitz false flag, the Nazi Party wanted to propagate the idea of German public support for Adolf Hitler. Nazi Party officials hosted a live German State Radio broadcast of a torchlit "parade" through Berlin. Because radio offered no visual evidence, listeners lacked the ability to hold broadcasters accountable. The size of the parade could be exaggerated by Nazi officials. Propagating this idea—that a majority of people supported Hitler—influenced some Germans into erroneously deciding Hitler was genuinely popular.

During Operation Hummingbird, Nazi Party operatives secretly murdered Ernst Rohm, Germany's head law enforcement official. Rohm was the equivalent of an FBI director. The idea behind Operation Hummingbird was to sell the lie, or propagate the idea, that the entire incident was an overthrow attempt by Rohm. It worked. During the operation, roughly one hundred Nazi Party enemies were murdered and replaced by Nazi Party allies.[40] By controlling Germany's law enforcement apparatus, the Nazi Party would be less accountable to the public.

Having seized majority control of the German legislative branch, Nazi Party officials then quietly utilized centralized legislative decision power to pass a law "concerning the Highest State Office of the Reich." This new social mechanism legislated that, upon the German president's death, the Office of the President and the Office of the Chancellor would automatically merge. The new law effectively rigged the decision as to who would become the next leader of Germany. Executive decision power was undemocratically being centralized in the Nazi Party.[41]

After the death of President Hindenburg, the Office of the President and the Office of the Chancellor automatically merged. Chancellor Hitler became the highest-ranking government official in Germany without receiving a single vote.[42]

1944: The Arm of the Invisible Government

Propaganda techniques have existed for millennia. Edward Bernays was an Austrian author who referred to propaganda as the "arm of the invisible government." This proved accurate in Nazi Germany, where Bernays' books were utilized by Nazi Party officials to create marketing campaigns that communicated to the German 99% the idea that Polish Jews were an imminent threat.[43]

Hitler described the practice very succinctly: "The most brilliant propagandist technique will yield no success unless one fundamental principle is borne in mind constantly and with unflagging attention. It must confine itself to a few points and repeat them over and over."[44]

Like playing a song over and over, this method of persuasion eventually results in the 99% singing along. Centralized control of communication systems is therefore a prerequisite for dragging along the 99%. Once communication power is centralized, the 1%, using a mockingbird media, can repeat a lie until it is accepted as truth.

At the same time, legislation, such as laws criminalizing the communication of any information legally classified as a "secret," enables the censorship of accountability information. Accountability power becomes centralized.[45, 46, 47]

CHAPTER 5

Secrecy Laws

1945: Ending World War II

In 1945, the U.S. government dropped two nuclear bombs on Japan: one in Hiroshima on the morning of August 5, and one in Nagasaki on the morning of August 6. Both bombs were dropped around 8:15 a.m. local time. Commuters were headed to work, and children were walking to school. According to eyewitnesses, there was a bluish light, like an electrical short, followed by tornado-like winds and black rain.[48]

Then everything went quiet. Victims wandered the streets unable to speak. Some of them had no hair or skin. A handful of local hospitals then overflowed with thousands of people experiencing uncontrollable vomiting and diarrhea. Over 100,000 people were killed between the two bombings.[49]

The U.S. 1% communicated to the U.S. 99% that World War II was ending because U.S. taxpayers had funded and developed nuclear weapons technology. But that idea was propaganda. Bombing Hiroshima and Nagasaki was unnecessary. As U.S. government officials would later confess, the bombings were instead intended to intimidate Russia.[50]

Secret U.S. government documents show that several offers of surrender had been sent from the Japanese emperor to U.S. government officials. In June of 1945, two months before the Hiroshima and Nagasaki bombings, William D. Leahy, chief of staff to U.S. President Harry Truman, documented his understanding in top secret government documents: "At the present time… a surrender of Japan can be arranged with terms that can be accepted by Japan and that will make fully satisfactory provision for America's defense against future trans-Pacific aggression."[51]

Leahy later blew the whistle on President Truman: "The use of this barbarous weapon at Hiroshima and Nagasaki was of no material assistance in our war against Japan."[52]

According to Army General Leslie Groves, director of the Manhattan Project, which covertly created the first U.S. nuclear bombs, "There was never any illusion on my part that Russia was our enemy, and that the project was conducted on that basis."[53]

After bombing Hiroshima, President Truman voiced his satisfaction with the "overwhelming success" of "the experiment."[54]

1946: Nuclear Proliferation

Nuclear technology was not an end but a beginning. By comparison, modern nuclear weapons (hydrogen bombs) are roughly a thousand times more powerful than either bomb dropped on Japan in 1945. At the time, American scientists theorized that fewer than 100 hydrogen bombs would be needed to end human life on Earth. For decades, that information was classified as a top secret and was unavailable to the U.S. taxpayers funding said bombs.[55]

Military-industrial corporations created a conflict of interest using a social mechanism involving bribery. They began trading option power (financial resources spun as campaign donations) for covert decision power (over congressional tax spending). Over the latter half of the twentieth century, U.S. politicians whose election campaigns were funded by military-industrial corporations decided to pour $5 trillion U.S. tax dollars into secretly constructing 4,000 hydrogen bombs.[56]

Mathematically speaking, nuclear war becomes more probable each time humans manufacture another nuclear bomb. If there are zero nuclear bombs, there is zero chance of nuclear war. Practically speaking, building nuclear bombs, using a near-endless supply of U.S. tax dollars, was simply highly profitable for military-industrial corporations.[57]

1947: Roswell

In the summer of 1947, an object crashed in Roswell, New Mexico, a town located 113 miles from the first U.S. nuclear test site. Local government officials immediately communicated to local newspaper reporters that they had recovered a "flying disc," implying it was extraterrestrial.[58]

Higher-ranking U.S. government officials wanted to propagate a different narrative. They wanted people to believe that extraterrestrials did not exist. They would need to conceal the evidence to avoid accountability to the public.[59] Furthermore, by propagating disinformation, U.S. government officials could corrupt evidence by contaminating true witness testimony with false witness testimony. Within twenty-four hours, U.S. government officials had successfully propagated a new story: the object was a weather balloon.[60]

U.S. government officials had a number of reasonable justifications to keep extraterrestrials a secret. For instance, public fear of extraterrestrial invasion or religious unrest could globally destabilize society. In other words, decision-makers feared the public would descend into chaos and anarchy if they knew the truth.

More importantly, U.S. government officials realized they could reverse engineer extraterrestrial technologies to create new weapons and commercial products. The financial benefits of those new technologies could be worth trillions of dollars to a nation's economy. Criminally punitive secrecy laws would be needed to ensure that national security information, be it military or economic, would not be leaked.

Shortly after Roswell, the U.S. military opened Projects Sign, Grudge, and Blue Book to investigate the extraterrestrial presence on Earth.[61] Those projects paralleled a U.S. government propaganda campaign intended to perpetually persuade the public that extraterrestrials do not exist.

The Foundation of Power: Information Resources

Information is arguably the most valuable resource on Earth. If you have information pertaining to the location of previously undiscovered resources—such as gold mines, water, or oil wells—you have the option power to capitalize said resources.

Similarly, having information resources pertaining to advanced technology, such as nuclear weapons, provides the option power to monetize it or leverage it during a war. Any nation acquiring that technology resource would want to keep it a secret from other nations. It could help win the next world war.

Protecting secrets was therefore a matter of national security, socially and economically, for every nation. Following World War II, the U.S. was looking to create a permanent social mechanism that could globally collect and analyze national security information (intelligence) during peacetime. If another nation developed technology resources on par with nuclear weapons, U.S. government officials wanted to know about it. As a result, the National Security Act was passed in 1947. The resulting social mechanism was the CIA.

The CIA was an entirely new agency designed to operate outside the three-branch accountability mechanism (checks and balances) set forth under Articles I–III of the U.S. Constitution. The CIA would not be a part of or accountable to the U.S. executive, legislative, or judicial branches.

The U.S. 1% had effectively created a fourth branch of government by propagating the concept of national security. Craving security in the wake of a horrific second world war, the U.S. 99% acquiesced. To the untrained eye, the CIA was an intelligence collection asset to the U.S. taxpayers funding it. In reality, the CIA was a threat to U.S. national security because the four powers were being centralized.

Information Governance

In 1949, two years after the CIA was established, the Central Intelligence Agency Act was passed. This law formally exempted the CIA from government accountability. All information involving tax dollars budgeted for the CIA was classified as top secret. By law, leaking top secret information would be punishable by death. This included any information involving covert operations, project names, or military-industrial contracts.

Accountability power over the CIA became fully centralized in the handful of men running the CIA. In other words, only the CIA could hold the CIA accountable. The CIA Act expressly violated the U.S. Constitution. Among other things, Article I empowers taxpayers with accountability over their tax dollars: "A regular Statement and Account of the Receipts and Expenditures of all public Money shall be published." 62, 63, 64, 65

The CIA was a black hole for public money. In the name of national security, a small, dark corner of the U.S. government had been disconnected from taxpayers. It was a slippery slope. The CIA was required to communicate with and be accountable to no one. Both powers were being undemocratically centralized in CIA officials.

CIA Operation Paperclip

World War II allowed covertly embedded German intelligence agents to obtain sensitive information involving both the U.S. and Russia. German intelligence agents were therefore empowered. Following the war, they possessed information resources sought by both the U.S. and Russia. German intelligence agents could avoid going to prison for war crimes by offering up various secrets quid pro quo. At the time, Russia was racing to keep up with the U.S. nuclear program.

Under Operation Paperclip, German intelligence agents with connections to Russia were secretly smuggled into the Americas by CIA agents. These German intelligence agents could then resume communicating with their colleagues embedded in Russia. By cooperating with the CIA, they had the option of living free in the Americas.

In September 1949, the fruits of Operation Paperclip were put to the test. Could the CIA see the future using covert intelligence-collection processes? CIA Director Roscoe Hillenkoetter was summoned to testify before the U.S. legislative branch regarding the Russian nuclear program. After consulting with the CIA officials handling Operation Paperclip, Hillenkoetter told the Congressional Committee on Atomic Energy, "The earliest possible date by which [Russia] might be expected to produce an atomic bomb is mid-1950, and the most probable date is mid-1953."[66] It was then discovered that Russia had already produced its first atomic bomb.[67]

Shell Corporations and Identity Fraud

To better verify the intelligence produced by Operation Paperclip, a new option was created: U-2 and SR-71 spy planes.

The U-2 could secretly take aerial photographs of Russia, allowing the CIA to estimate the number of Russian nuclear missiles. Because it was capable of flying at 80,000 feet at the edge of Earth's atmosphere, the Russian military could hardly detect the U-2 spy plane, let alone shoot it down. Russia was now inadvertently communicating with the U.S. by leaving its nuclear weapons in plain view. Top secret spy plane technology was connecting the CIA to information resources.[68]

The SR-71 was another flying camera. Also flying at 80,000 feet, it could photograph not just large missiles but also words on documents. It traveled at Mach 3, three times the speed of sound. Reaching temperatures of 450° F, the SR-71 needed to be built using titanium. Ironically, the only country with substantial titanium resources was Russia.[69]

To access that titanium, the CIA created bogus businesses, allowing the agency to purchase the titanium without Russians recognizing the true buyers. Fraudulent identifications, such as fake international passports, were used to set up shell corporations. Using these fraudulent identifications, CIA agents could now unilaterally commit untraceable crimes.[70,71]

The U.S. Military-Industrial Complex

As it turned out, German war criminals had been lying to the CIA. The former head of Nazi intelligence, Reinhard Gehlen, was at the apex of Operation Paperclip. Attempting to appear valuable in the eyes of the CIA and avoid prosecution for war crimes, Gehlen fabricated accountability information by exaggerating the number of Russian nuclear weapons.

Once U.S. military officials realized that Gehlen was lying, they did nothing. U.S. military officials were the middlemen between Gehlen and the U.S. legislative branch. The U.S. military budget was driven by Gehlen's numbers. Lying to congress was profitable. U.S. military officials had numerous conflicts of interest: (1) their salaries were increasing through expansion-necessitated promotions, (2) they could obtain higher-paying jobs in the future with U.S. military-industrial corporations, and (3) they had insider information on U.S. government contracts and could purchase stock in U.S. military-industrial corporations.[72, 73]

Similarly, in exchange for financial resources (campaign donations), U.S. politicians continued promising future tax dollars to these military-industrial corporations. U.S. President Dwight Eisenhower described it thusly in 1961: "In the councils of government, we must guard against the acquisition of unwarranted influence, whether sought or unsought, by the military-industrial complex. The potential for the disastrous rise of misplaced power exists and will persist. We must never let the weight of this combination endanger our liberties or democratic processes."[74]

The U.S. military-industrial complex discovered how to undemocratize the U.S. by centralizing decision power in itself. The key social mechanism was personal financial incentives. Corporations were baiting the hook to suit the fish. They used personal financial incentives to create irresistible conflicts of interest for both political and corporate decision-makers. Simply put, almost every decision-maker was profiting.

The Fear-Based Logic of Preemptive Strike Theory

Operation Paperclip had some U.S. politicians on the brink of invading Russia under preemptive strike theory. Unlike the doctrine of self-defense, where Entity X attacks Entity Y because Entity Y is currently attacking Entity X, the doctrine of preemptive strike attempts to see the future. Prior to starting World War II, preemptive strike theory was part of the logic used to justify Japan's false flag in Manchuria and Germany's false flag in Gleiwitz.

Under the fear-based logic of preemptive strike theory, Entity X attacks Entity Y to preempt and prevent Entity Y from attacking Entity X in the future. It assumes a threat will inevitably manifest. There are two problems with this line of thinking. The first problem is that humans cannot see into the future. The second problem is that the only evidentiary requirement for a preemptive strike is a perceived threat. As history has proven, a threat can be covertly manufactured by those with centralized power through lies, false flags, and other propaganda.

China was a "threat" to Japan, Poland was a "threat" to Germany, and Russia was a "threat" to the U.S. In all three instances, an attack was not actual or imminent. Japan, Germany, and the U.S., respectively, manufactured imminent threats in their minds. While war was profitable, decisions were also genuinely driven by fear.

In the U.S., CIA officials feared a third world war. According to its secret charter, the CIA could preemptively strike any perceived threat. Strike techniques included propaganda, economic warfare, and preventive direct action—and U.S. taxpayers would unknowingly fund it. Passive intelligence collection was not enough. The CIA was empowered. CIA officials were undemocratically sanctioned to secretly manipulate any government in the world.[75]

CHAPTER 6

Mass Media

CIA Operation Mockingbird

After studying Nazi propaganda techniques, the CIA wanted to go one step further. CIA officials were interested in secretly centralizing communication power globally. Under Operation Mockingbird, hundreds of CIA agents were tasked with covertly infiltrating mass media outlets in every country. The CIA could then censor independent investigative journalists. Like mockingbirds, international media outlets would then undemocratically sing the songs chosen by CIA officials. As CIA agent William Bader described it, "You don't need to manipulate *Time* magazine, for example, because there are [CIA] people at the management level."[76]

The CIA also created a database called Propaganda Assets Inventory (PAI). PAI contained the names and contact information for CIA media assets serving as television executives, radio personalities, journalists, and book publishers at employers such as Time, ABC,

NBC, CBS, Newsweek, National Enquirer, Associated Press, Reuters International, United Press International, U.S. News and World Report, the Washington Post, and the New York Times. The key was creating conflicts of interest for decision-makers. Using U.S. taxpayer-funded personal financial incentives and a patriotic spin, media executives were bribed and persuaded to propagate a CIA-controlled narrative.[77]

1948: The Secret CIA Overthrow of Italy

In 1948 as part of Operation Mockingbird, the CIA covertly directed a popular magazine to launch a propaganda campaign supporting a CIA-friendly Italian political candidate, Alcide de Gasperi. De Gasperi was placed on the cover of the April 1948 issue of *Time* magazine. Campaign resources were covertly diverted to de Gasperi by CIA agents. According to CIA agent F. Mark Wyatt, "bags of money" were given to de Gasperi and other U.S.-friendly Italian politicians.[78]

The Italian 99% lacked accountability power and were unable to recognize that de Gasperi was receiving "campaign donations" from CIA agents. Similarly, the U.S. 99% lacked accountability power and were unaware their tax dollars were being used to rig democratic elections overseas. The CIA hoped it could preemptively strike other nations by rigging their elections and preventing the rise of another Benito Mussolini or Adolf Hitler.[79, 80]

De Gasperi subsequently won the election and became head of the Italian government. Believing they were democratically choosing their Prime Minister, the Italian 99% were clueless to the fact that de Gasperi was secretly controlled by the CIA. The overthrow of Italy was a confidence builder for the newly formed CIA.[81]

Connecting the U.S. to the Middle East: Israel

In 1948 U.S. taxpayer funding helped undemocratically establish the State of Israel in the middle of Palestine. From the perspective of Arabs, it was the equivalent of Arabs establishing a new Islamic State in the middle of Texas. In reality, U.S. government officials were strategically positioning themselves to steal oil and other natural resources from Arabs.

Regardless of Christian, Jewish, Islamic, or any other religious theories as to who is entitled to live in Palestine, one simple fact remained: there were already Arabs living in Palestine in 1948. The U.S.-controlled United Nations decided to partition Palestine. Half would be a Jewish state (Israel), and half would remain an Arab state. The capital city, Jerusalem, would be shared. Indigenous Arabs in the new Jewish state were kicked out of their homes.[82]

At the time, the U.S. government had the military and nuclear firepower to take any land resources it wanted. Not even Russia objected. The U.S. 1% was preaching democracy but practicing conquest. Critical natural resources, like oil, were covertly being surveyed and acquired. Clark Clifford, one of President Truman's advisors, spun the U.S. creation of Israel as the stabilization of an "unstable" Middle East:

> "I fully understand and agree that vital [U.S.] national interests are involved. In an area as unstable as the Middle East, where there is not now and never has been any tradition of democratic government, it is important for the long-range security of our country, and indeed the world, that a nation committed to the democratic system be established there, one on which we can rely. The new Jewish state can be such a place. We should strengthen it in its infancy by prompt recognition."[83]

U.S. national security interests clearly conflicted with global public interest. U.S. politicians realized that the entire U.S. military was powered by oil. Every plane, tank, and boat needed oil. U.S. oil corporations had wrapped their tentacles around the U.S. military to such a degree that oil became a matter of national security. Instead of deciding to create alternative energy options, U.S. government officials decided to create Israel.

As U.S. Senator Jesse Helms later confessed, Israel was "America's aircraft carrier in the Middle East."[84]

The overall spin was that Israel was "established," not stolen, and that Israel was necessary for the "containment of communist Russia."[85]

It was a turning point for the U.S. government. Taking land from Arabs was an abandonment of democratic principles. Democracy requires equal protection. It requires that the strong protect the weak rather than attack them or steal their resources. Aggression, greed, and fear were taking priority over equality, communication, and accountability. The decisions of the U.S. government sent a very undemocratic message to Arabs: non-Americans are unequal to Americans.[86, 87, 88, 89]

1949: The Secret CIA Overthrow of Syria

In 1949 U.S. taxpayers unknowingly funded a CIA overthrow of Syria. Adib al-Shishakli was covertly installed as a puppet leader, creating a military dictatorship. Soon after, the Syrian 99% regained control of their government and interrogated CIA-corrupted Syrian politicians. Those politicians then went on Middle East television networks and communicated they had taken money from the "sinister Americans." Syrian politicians were told they would be assassinated if they did not cooperate with the CIA.[90]

The U.S. embassy in Syria was subsequently surrounded by Syrian soldiers. CIA Agent Howard Stone, who was covertly operating out of the U.S. embassy as a diplomat, emerged and confessed the CIA's guilt.[91]

In the wake of World War II, most countries were attempting to rebuild trust and improve communications. A great deal of the trust the U.S. accrued during the world wars was instantly lost after Syria. Using new television technology, the U.S. was held accountable—locally.[92] The Syrian 99% communicated evidence to other Middle East countries that the U.S. government was disrespecting Syria's sovereignty through aggressive covert action. Halfway around the world, the CIA-controlled U.S. media made little mention of the incident.[93] Again, the U.S. 99% were clueless.

The first power of government is communication. For purposes of global democracy, every nation must communicate with its own people as well as with every other nation. To ensure international communication, every nation should have an embassy in every other nation. Embassies create a global communications network. They can prevent world wars. They allow world leaders to communicate with foreign ambassadors face-to-face in real time.

Every time CIA officials leverage a U.S. embassy as a base for covert action, they undermine democracy and compromise U.S. national security, as well as global security. Agent Stone's confession provided evidence that U.S. embassies were being utilized as safe houses for undercover CIA agents. After the Syria debacle, every nation would have to question the wisdom of allowing a U.S. embassy on their soil.

1950: The War to Resist U.S. Aggression (The Korean War)

At the end of World War II, Japan agreed to relinquish control of the Korean State. The U.S. and Russia, fresh from victory, were dividing the spoils. Koreans, like Arabs, lacked the military firepower to resist the might of these two powerful nations. Roughly speaking, the U.S. took the southern part of Korea, and Russia took the northern part.[94] In theory, the two superpowers would then go their separate ways.

Rather than respecting a unified Korea and honestly facilitating democratic elections, the U.S. and Russia patiently tried to impose their separate agendas of conquest. The U.S. 1% accused Russians in the north of socialist oppression. The Russian 1% accused Americans in the south of capitalist oppression. From the perspective of the Korean 99%, both the U.S. and Russia were mirror images of each other, propagating lies and oppressively attempting to control Korea's resources.[95]

The political spin continued to escalate as both the U.S. and Russia renamed their respective regions. The south was formally renamed the Republic of Korea, and the north was formally renamed the Democratic People's Republic of Korea. CIA and Russian KGB agents were immediately meddling in Korea's elections.[96]

In 1950 North Korea allegedly invaded South Korea, but not every nation spun it as the "Korean War." In China, for example, the Korean War was called the "War to Resist U.S. Aggression and Aid Korea."[97]

Secret CIA Drug Cartels

Coinciding with the Korean War was a less-visible propaganda campaign by CIA officials stationed in South Korea. U.S. government decision-makers were relying on CIA intelligence to direct the U.S. troops on the ground. With little accountability, CIA field agents had been falsely reporting one successful CIA operation after another.[98, 99]

Suspicious of the unusual success rate, Loftus Becker, the CIA deputy director of intelligence, was deployed to South Korea to investigate. Becker reported that most of the information communicated from CIA agents in Seoul was false or misleading. Becker added that the CIA's capabilities in the Far East were hopeless and then submitted his resignation from the CIA. News of the incident never reached U.S. taxpayers because the CIA controlled the U.S. media.[100, 101]

According to Becker, U.S. taxpayers were funding the presence of roughly 200 CIA agents in Seoul, and none of them spoke Korean. In addition, millions of U.S. tax dollars had been covertly diverted to a global heroin cartel based out of Burma, a secret source of profit for some CIA officials. As an accountability-free agency, CIA officials were able to empower themselves as dealers in global resources, both legal and illegal. CIA drug cartels were "black" operations. They were invisible. There were no paper trails. CIA officials had created a new source of cash flow separate from congressionally approved tax dollars.[102]

If U.S. legislators ever reduced CIA congressional funding, the CIA would keep going strong.

1951: The Secret CIA Overthrow of Iran

During the late 1940s, U.S. presidents had met with several Middle East kings, offering U.S. military security in exchange for providing discounted oil to U.S. corporations. Many kings capitalized on the opportunity to prostitute the U.S. military in exchange for oil. While the kings skimmed oil revenues, a majority of their people lived in poverty.[103] Middle Eastern protestors were violently silenced by local security forces trained and armed by the CIA.

In 1951 Iran's democratically elected National Front Party took over majority control of the Iranian senate. The Iranian senate had previously been controlled by the king of Iran (the shah). The National Front Party proceeded to expel foreign oil corporations from Iran. In 1953 the CIA secretly funded a violent overthrow of Iran and its democratically elected prime minister, Mohammad Mosaddegh. The overthrow enabled the shah's return to power. In return, the shah transferred roughly 40% of Iranian oil interests to U.S. corporations.[104]

The Iranian 99% had democratically spoken during the 1951 elections, yet here was the CIA, secretly disrespecting the very democratic process it publicly championed. The CIA had effectively snuck into the room with a briefcase full of U.S. tax dollars and overthrown the Republic of Iran by reinstalling a king. The National Front Party was not the Nazi Party. The CIA was not preventing World War III. The Iranian people fervently supported the National Front Party's attempts to take public ownership of Iran's resources, such as oil, and liberate Iran from foreign control.

Instead, Iran returned to a state of violent oppression and poverty. Under Operation Ajax, the CIA supplied the weapons and funding needed to secure the shah's power. By overthrowing Iran, CIA officials were in no way protecting democratic processes—they were protecting their own financial interests as well as the U.S. military's dependence on oil.[105]

1952: The Secret CIA Overthrow of Cuba

In 1952 U.S. taxpayers unknowingly funded a CIA overthrow of Cuba. Four years earlier, the Cuban 99% had democratically elected Carlos Prío Socarrás as president. Prío subsequently lobbied Central American countries to work together to combat threats such as the financial interests of the U.S. 1%. In response, the CIA covertly overthrew Prío using Cuban military personnel, including General Fulgencio Batista. Batista, who had spent the previous eight years preparing for the operation in New York and Florida, seized Cuba's military, police, television studios, and radio stations.[106, 107]

Just like the shah in Iran, Batista had a conflict of interest. His personal financial interests conflicted with Cuba's national security interests. Batista's decisions financially benefited him but not his country. He abused his power by aligning Cuban resources with U.S. corporate interests, including mining, oil, sugar, and banking. Over 20,000 political activists were murdered by CIA-trained Cuban security personnel as the country descended into poverty.[108, 109]

By 1958 U.S. corporations owned roughly 90% of Cuban mines. That same year, however, Cuban activists overthrew Batista, placing attorney Fidel Castro in power. Castro implemented free healthcare for all Cubans, kicked out U.S. corporations, and retook control of Cuba's resources. With the assistance of the CIA, Batista fled Cuba for Portugal. At the time, Batista's net worth was estimated at over $300 million.[110]

1953: CIA Project MK-ULTRA

In 1953 CIA scientist Frank Olson internally blew the whistle on the U.S. government's use of biological weapons. He subsequently "fell" from the thirteenth-floor window of a New York hotel room.[111] Olson had been drugged with LSD by CIA Agent Sidney Gottlieb, the well-paid head of Project MK-ULTRA.[112] Olson's death was consistent with the CIA's secret assassination manual, which states, "No assassination instructions should ever be written or recorded.... The most efficient accident, in simple assassination, is a fall of seventy-five feet or more onto a hard surface."[113]

Under Project MK-ULTRA, Gottlieb secretly conducted mind-control experiments on innocent Americans. CIA agents were dispatched to more than thirty U.S. colleges where they drugged unsuspecting students with LSD, marijuana, cocaine, PCP, heroin, and Rohypnol (the "date rape drug"). Female students were then followed by male CIA agents, allegedly to study their behavior.[114] Using taxpayer funding, the documented goal of the CIA agents involved was to answer the question, "Can we get control of an individual to the point where [he or she] will do our bidding against [his or her] will?"[115]

The CIA was attempting to seize decision power without consent. On a grander scale, CIA officials were interested in using drugs, for example, to brainwash a Middle Eastern student attending a U.S. college. The student would be programmed to return home and shoot a foreign leader, such as a king.[116]

Under Project MK-ULTRA, several pharmaceutical corporations and healthcare providers allowed the CIA secret access to patients. For example, using drugs of interest to the CIA, Dr. Donald Ewen Cameron was paid to induce comas in patients under his care. In one instance, a comatose patient was forced to listen to one recorded message, in total darkness, for more than 100 days straight.[117, 118]

To blackmail the CIA for additional U.S. taxpayer funding, Dr. Cameron secretly recorded adolescent patients performing sexual acts on CIA officials. Left unaccounted for, CIA officials were proving capable of horrific acts.[119, 120]

1954: The Secret CIA Overthrow of Guatemala

In 1954 U.S. taxpayers unknowingly funded a CIA overthrow of Guatemala. Years earlier a U.S. corporation called United Fruit Company (a.k.a. Chiquita Brands International) had taken ownership of more than one-third of the land in Guatemala. Through banana exports alone, United Fruit Company was bringing in more revenue than the entire Guatemalan government.[121] In response, the Guatemalan legislature passed land reform laws. Under those laws, over half of the United Fruit Company–controlled land was taken back. United Fruit Company was compensated through IOUs in the form of Guatemalan government bonds.[122]

At the time, CIA Director Allen Dulles was a shareholder in United Fruit Company. Dulles covertly launched Operation PBFORTUNE— a plan to secretly overthrow Guatemala's democratically elected government. After a CIA agent left plans for the overthrow in a Guatemalan hotel room, the conspiracy was communicated to the Guatemalan 99% by newspaper. Guatemalan laborers took up arms as word spread of an attempt by the U.S. to create a "banana republic" in Guatemala. The CIA subsequently abandoned the operation.[123] Dulles's personal financial interest in United Fruit Company is a textbook example of a conflict of interest. He was prioritizing personal profit over democracy.

Soon after, Dulles tried again, launching Operation PBSUCCESS. Using U.S. tax dollars, the CIA secretly provided weapons, training, and promises of power to former Guatemalan soldiers, including Carlos Castillo Armas. CIA propaganda assets, such as a local radio station called the Voice of Liberation, made the Guatemalan 99% fearful of resisting the U.S. military and the Castillo Armas overthrow. After violently taking power, Castillo Armas repealed the land reform laws, and United Fruit Company regained control of Guatemalan land.[124] Over the next thirty years, roughly 200,000 Guatemalans would be killed resisting United Fruit Company and a Guatemalan government supplied with weapons from the CIA.[125]

1955: The Vietnam War, Southeast Asia

By 1955 the CIA was developing a playbook. When a U.S. corporation sought to take control of resources in a foreign country, the CIA would secretly step in.

Using "economic hitmen," the CIA would first encourage heads of state to peacefully cooperate with U.S. corporations. If a head of state refused, the CIA would covertly orchestrate an overthrow by bribing military officials or politicians. A puppet leader was typically installed through legislation, election rigging, intimidation, or assassination.

As detailed in books like *Confessions of an Economic Hitman*, CIA officials were clearly comfortable assassinating democratically elected officials. The CIA would install puppet leaders who epitomized greed and violence, as evidenced by the bribes and weapons they accepted from the CIA. Most of the countries in the South Pacific fell victim.

- *Indonesia:* The CIA secretly used U.S. tax dollars to fund a military overthrow of democratically elected President Sukarno. The overthrow was spun as the "Thirtieth of September Movement." Under Operation HAIK, Indonesian Army General Suharto was installed as president. During General Suharto's regime, roughly one million protestors were killed by U.S. taxpayer-funded security forces. U.S. tax dollars ensured the repeated reelection of General Suharto, who embezzled over $10 billion during his thirty years in office.[126]
- *Philippines:* The CIA secretly used U.S. tax dollars to rig elections, ensuring the repeated reelection of President Ferdinand Marcos. During his time in power, Marcos embezzled over $4 billion.[127] As a result of U.S. corporate

infrastructure projects, the Philippine national debt exploded from $2 billion to $25 billion.[128] The CIA is also linked to the deadly 1957 plane crash of President Ramon Magsaysay and the 1983 assassination of presidential candidate Benigno Aquino Jr.[129, 130]

- *Cambodia:* The CIA secretly used U.S. tax dollars to overthrow the Cambodian government and install Lon Nol as president. Nol immediately sent reluctant Cambodian troops to fight alongside the U.S. in North Vietnam. Over one million war protestors were subsequently killed by CIA-backed Cambodian regimes. To save Nol from his own people, the U.S. government relocated him from Cambodia to Fullerton, California.[131, 132]
- *Laos:* The CIA secretly used U.S. tax dollars to create opium cartels and fund black operations. Under Operation Momentum, the CIA recruited Laotian tribes to fight communists and encouraged the tribes to grow opium instead of rice. The tribes then became dependent on CIA air drops for food. Tribes who refused to fight alongside CIA-controlled armies or to grow opium were cut off from food supplies and were bombed. Roughly 10% of the Laotian people were killed, while approximately 25% became refugees.[133, 134, 135, 136]
- *Thailand:* The CIA secretly used U.S. tax dollars to assassinate the Thai king. At the close of World War II, Thailand's twenty-year-old King Ananda Mahidol was "accidentally" shot

in the forehead with a Colt .45. The gun had been gifted to King Ananda by the head of the *Bangkok Post* newspaper, Alexander MacDonald, a CIA asset. Coincidentally, King Ananda's successor, Bhumibol Adulyadej, was the last person to see King Ananda alive.[137] Over a seventy-year reign, King Bhumibol used U.S. taxpayer-funded weapons to fight off democratic uprisings while accruing a net worth of roughly $30 billion.[138, 139]

- *Taiwan:* The CIA secretly used U.S. tax dollars to install Sun Li-jen as leader. The overthrow was thwarted when Russian KGB-trained Chiang Ching-kuo, the head of the secret police, had Sun arrested.[140, 141]

Southeast Asia was a geopolitical gateway to Russia. Just like in Korea, South Vietnam was controlled by the U.S., and North Vietnam was controlled by Russia. The U.S. and Russia were again mirror images of each other. Regardless of which side prevailed, Vietnamese resources would be pilfered by the Vietnamese 1% and the foreign entities that controlled them.

1956: Covert Control of Central and South American Resources

Similar to its efforts in Southeast Asia, the CIA was meddling in Central and South America.

- *Panama:* The CIA secretly used U.S. tax dollars to fund a violent military overthrow of the

Panamanian government, installing General Manuel Noriega as president. In exchange for Noriega's tripling the size of the Panamanian military and purchasing weapons from the U.S., the CIA protected him, Pablo Escobar, and the Medellin drug cartel.[142] Once Noriega became uncooperative in the late 1980s, the U.S. invaded Panama and removed Noriega.[143, 144]

- *Honduras:* The CIA secretly used U.S. tax dollars to fund a violent military overthrow of the Honduran government, installing the U.S.-friendly General Policarpo Juan Paz García as president. General Paz and other Honduran government officials helped the CIA smuggle weapons into South America in exchange for safely smuggling drugs into the U.S. and other countries.[145, 146]
- *Brazil:* The CIA secretly used U.S. tax dollars to fund a violent military overthrow of the Brazilian government. President João Belchior Marquis Goulart had proposed expanding voting rights and kicking foreign corporations out of the country. The CIA then installed the U.S.-friendly Humberto de Alencar Castelo Branco as president. Castelo Branco immediately brought on crippling debt by opening Brazil to the International Monetary Fund (IMF) and the World Bank, which sold Brazil the loans needed for U.S. corporate infrastructure projects.[147, 148]
- *El Salvador:* The CIA secretly used U.S. tax dollars to rig the 1984 El Salvador presidential

election, installing José Napoleón Duarte as president. Duarte's CIA-funded death squads killed roughly 50,000 dissidents.[149, 150]

- *Argentina:* The CIA secretly used U.S. tax dollars to back Operation Condor. Approximately 30,000 Argentine activists were killed as a result of CIA support.[151, 152]
- *Bolivia:* The CIA secretly used U.S. tax dollars to fund a violent military overthrow of the Bolivian government, installing the U.S.-friendly General René Barrientos Ortuño as president. Barrientos then amended the Bolivian Constitution to permit his CIA-funded reelection. Two presidential terms later, following the democratic election of "leftist" Juan José Torres González, a CIA-backed military coup installed General Hugo Banzer Suárez as president. Banzer subsequently banned all "left" political parties and killed more than 200 political dissidents. When Banzer's second term ended, the CIA secretly poured in more U.S. tax dollars to ensure the election of General Juan Pereda Asbún, who apparently received 200,000 more votes than the actual number of registered voters.[153, 154]
- *Columbia:* After U.S.-based Occidental Petroleum Corporation discovered the Caño Limón oil field in Columbia, the CIA moved in. Since 1986 more than 4,000 leftist politicians have been assassinated by CIA-backed Columbian death squads.[155, 156]

- *Nicaragua:* The CIA secretly used U.S. tax dollars to fund a violent overthrow of Nicaragua. The CIA hired mercenaries, but rather than call them terrorists, the CIA spun them as "contra rebels." The operation was eventually revealed by an Iranian whistleblower. Republican President Ronald Reagan was subsequently implicated in the Iran-Contra Affair. Despite extraordinary evidence to the contrary, a House Congressional committee, led by Republican House Representative Dick Cheney, did not hold Reagan accountable. Ten years later, Republican President George H.W. Bush pardoned every U.S. government official charged in the Iran-Contra Affair, all of whom were Republicans.[157, 158]

When it served personal and political party interests, U.S. executive branch officials were abandoning democratic principles.

Subhuman Resources

In Southeast Asia, Central America, and South America, the U.S. was not only capitalizing natural resources but also human resources. For example, foreign puppet leaders would decide to hire U.S. corporations instead of local businesses to complete infrastructure projects such as roads, power plants, sanitation, manufacturing, oil refining, and mining. High-paying jobs were fulfilled by Americans while high-risk jobs were left for the local population.

To visiting Americans, the local populations were often perceived as subhuman. As described in *Confessions of an Economic Hitman*, the locals could be seen bathing and defecating alongside each other in a river.[159] The contrast in ways of life between Americans and non-Americans was unforgettable. In their own minds, CIA officials could rationalize their bad behavior by spinning it as modernizing third world countries.

Because the CIA controlled legislators in those countries, there were no minimum wage laws. Laborers could be paid less than a dollar a day for a seven-day work week. There were no child labor laws. Children as young as five years old could be forced into labor by impoverished parents. There were no maternity leave laws. Pregnant women would keep working or lose their jobs. The arrangement was ideal for maximizing profit.

1957: The Secret CIA Overthrow of Africa

In 1957 U.S. taxpayers unknowingly funded a secret CIA overthrow of Ghana. Ghana had declared its independence from British corporations, and the event was celebrated by some as a breakthrough for democracy in Africa. As President Truman put it ten years earlier, the people of Ghana were trying to "work out their own destinies in their own way."

Ghana's first democratically elected leader, President Francis Kwame Nkrumah, a U.S.-educated economist, declared, "We have a duty to prove to the world that Africans can conduct their own affairs with efficiency and tolerance and through the exercise of democracy. We must set an example to all Africa."[160]

Nkrumah introduced free education for all. He understood that education was inseparable from democracy. Education would facilitate communication and empower people with accountability information. Nkrumah also introduced free healthcare for all. As an economist, he understood the long-term cost and hygiene benefits of proactively providing healthcare resources.

Under the CIA's Operation Cold Chop, the Nkrumah administration was overthrown. General Joseph Ankra, the head of Ghana's National Investment Bank, was installed as head of state. Ankra immediately handed over control of Ghana's economy to the IMF and World Bank. Ghana subsequently descended back into a pre-Nkrumah state of poverty and human rights violations.[161, 162, 163]

The four powers dictate that every democracy include the right to accountability. The right to accountability necessarily involves the right to education and the right to healthcare. You cannot have an effective democracy if everyone is uneducated and unhealthy. Both of these rights are directly at odds with the modern centralized systems of for-profit higher education and healthcare. The CIA was exterminating the most democratic world leaders in order to centralize resources in a global 1%.

1958: The Secret CIA Overthrow of Europe

Immediately following World War II, as part of Operation Gladio, the U.S. left "stay-behind armies" in every possible European country. Those sleeper agents would set up secret bases, communications systems, weapons caches, and escape routes to prepare for any threat of Russian invasion. According to European media, Operation Gladio was "the best-kept, and most damaging, political-military secret since World War II."[164]

Operation Gladio instead gave the CIA the opportunity to overthrow every European country that failed to cooperate with the financial interests of the U.S.-led North Atlantic Treaty Organization (NATO). Gladio also provided the infrastructure for CIA heroin cartels in Europe. The following European countries are now known to have been impacted by Operation Gladio.

- *Austria:* The CIA secretly used U.S. tax dollars to create the Austrian Association of Hiking, Sports and Society. Members were secretly trained in weapons and plastic explosives. Following the discovery of secret CIA armies and weapons caches all over Europe, Austria demanded an explanation from U.S. President Bill Clinton. He refused. Nick Burns, the U.S. Ambassador to NATO, later stated, "The aim was noble, the aim was correct, to try to help Austria if it was under occupation. What went wrong is that successive Washington administrations simply decided not to talk to the Austrian government about it."[165, 166]
- *France:* The CIA secretly used U.S. tax dollars to covertly overthrow French President Charles de Gaulle. De Gaulle, however, still controlled

French communication systems. Through television and transistor radio technology, de Gaulle immediately responded by broadcasting a passionate plea for the French people to resist the overthrow: "Frenchwomen, Frenchmen, help me!" The resistance succeeded and is referred to in France as the Battle of the Transistors. De Gaulle demonstrated that being armed with communications systems can be as valuable as being armed with guns. As a result of their inability to trust the U.S., France and other nations began trading in their U.S. dollars for gold. To stop the out-flux of gold, U.S. President Richard Nixon eventually removed the dollar from the gold standard.[167, 168, 169]

- *Portugal:* The CIA secretly used U.S. tax dollars to create Aginter Press, a media agency that trained U.S.-friendly locals in terrorist techniques, including clandestine communication, infiltration, subversion, explosives, and assassinations.[170, 171]
- *Italy:* The CIA secretly used U.S. tax dollars to pay over $22 million in bribes to Italian politicians, including Italian Prime Minister Mariano Rumor. Rumor's successor, Aldo Moro, subsequently proposed "national solidarity" between his political party, the Christian Democratic Party, and the Italian Communist Party. Soon after, Moro was kidnapped and shot in the head by a group claiming to be the pro-communist Red Brigade. The Red Brigade was eventually exposed as a false flag,

a product of the Italian secret intelligence service (SISMI) and the CIA.[172, 173, 174, 175]

- *Ireland:* The CIA secretly used U.S. tax dollars to supply weapons to the Irish Republican Army (IRA). At his criminal trial, defendant Michael Flannery stated, "We were doing no wrong—we were working with the CIA." After mysteriously being acquitted, Flannery was named the grand marshal of the New York City St. Patrick's Day Parade.[176, 177]
- *Turkey:* The CIA secretly used U.S. tax dollars to overthrow Turkey in 1960, 1971, 1980, 1997, and 2016. Turkey, which borders Iraq, Iran, and Syria, has obvious strategic value to U.S. interests and subsequently receives aid in the form of hundreds of millions of U.S. tax dollars annually.[178, 179, 180, 181]
- *Switzerland:* The CIA secretly used U.S. tax dollars to fund Project 26. Project 26 was a secret U.S. Army left behind in Switzerland following World War II. It was mandated to resist a Soviet invasion and become active in the event the "left" achieved parliamentary majority.[182, 183]

In the mind of CIA officials, any left-leaning politicians were a threat to U.S. national security. This policy would come to a head when a highly intelligent and charismatic leftist became the U.S. President in 1960.

The Preferred Narrative: Fighting Communism

It's human nature to look for evidence that fits your preferred narrative. CIA Agent Ralph McGehee spent twenty-five years in the CIA, mostly in Southeast Asia. A former recipient of the CIA's Career Intelligence Medal, McGehee initially bought into the CIA's publicly stated narrative of "fighting communism." Eventually, he saw through the propaganda. According to McGehee, agents find the information they are ordered to find rather than finding the truth:

> "Sam Adams, a CIA analyst who fought the CIA at every step, said the Agency, in undercounting [enemy soldiers], refused to use what should have been its primary source—captured enemy documents. In my own experience, I discovered that the CIA buried any information that did not support its pro-war policies. Asian communist leaders set forth in their writings the plans and programs of their revolutions, but I doubt if any Agency operative, or any member of the best and brightest, ever read or, if so, understood those writings. The CIA recruited paid agents to tell the CIA what it wanted to hear, ignoring the mass of overt information that so disproved our rationales for the war. This practice epitomizes Agency operations from the beginning to the present."[184]

In other words, CIA officials find information that fits their preferred narrative. U.S. President Dwight

Eisenhower was told by his advisors that the CIA was incapable of making objective appraisals of its own intelligence information. As CIA analyst Abbot Smith described it, "We had constructed for ourselves a picture of [Russia], and whatever happened had to be made to fit into this picture."[185]

Justifications for the Vietnam War had been propagated using generic words such as "communism" and "leftists." There was a lack of accountability. If evidence of bad behavior by a foreign entity did not exist, it could be fabricated. Without accountability, the CIA could manufacture national security intelligence regarding communist threats and then package it, through a CIA-controlled mockingbird media, for the U.S. 99% to consume. At the same time, the perception of the world, for the Russian 99%, was similarly being driven by the Russian 1% through the KGB.[186, 187, 188]

U.S.–Russia Propaganda: Cold War vs. Hot War

The beauty of the Cold War between the U.S. and Russia was that there was no war at all. This was spin. It was certainly not a "hot war": there was no combat zone. The U.S. and Russia had been allies in World War II. Were these two countries a threat to each other? Of course, but every country can be seen as a threat to another. For that matter, every person is a threat to every other person. The problem with this line of thinking is that it's based on fear. It is a hallmark of national security propaganda. For a person who lives in fear, death is around every corner. No one else can be trusted. No one else should be empowered.

When fear dictates, the human mind preemptively strikes by proverbially charging around the corner to confront death. It is paranoid and maniacal. Yet this is what the U.S. and Russia were doing at a global level through the CIA and KGB. When individuals, instead of 100% of the group, are vested with power, childish fistfights between world leaders are inevitable. It's easy for seventy-year-old politicians to talk a big game when twenty-year-old soldiers have to back it up.

During the Cold War, the Russian 1% and the U.S. 1%, through repeated communications and a mockingbird media, genuinely terrorized the global 99%. The Cold War, however, was only a mental construct. Attack was not imminent. Nothing was "hot." This mental construct helped the U.S. 1% circumvent the U.S. Constitution, which prohibits the perpetual "raising" of the U.S. military. The founders of the U.S. understood that war only leads to more war. As the saying goes, violence begets violence.

The authors of the U.S. Constitution mandated that armies only be funded for two years at a time, not indefinitely. This clause was intended to keep the U.S. from repeatedly "charging around the corner" like a maniac. Under Article I of the Constitution, "The Congress shall have Power… To raise and support Armies, but no Appropriation of Money to that Use shall be for a longer Term than two Years."

The Cold War lasted more than forty years. There had to be a war for the military-industrial spending sprees to continue. Tax dollars would be needed to manufacture infinite numbers of tanks, planes, and nuclear bombs. It was in the best financial interests for both the U.S. 1% and the Russian 1% to keep spending.[189, 190]

1959: CIA Operation Mongoose

Just as the CIA was covertly overthrowing foreign governments, so too was the KGB. In 1959, Fidel Castro, a KGB-funded Cuban, overthrew the CIA-controlled Cuban government and undemocratically assumed the role of prime minister.[191]

Castro then passed land reform laws that transferred ownership of oil, sugar, coffee, and banking from U.S. corporations back to Cuba. Non-Cubans were banned from owning land. This did not go over well with U.S. government officials who were horrified Russia had control of a country situated only 100 miles from the U.S. coastline.[192]

In response, through Operation Mongoose, the CIA began a long string of assassination attempts on Castro. These efforts included lining Castro's scuba-diving suit with a deadly fungus, mixing a poison into his face cream, providing him with exploding and poisoned cigars, and giving his lovers poison pills to feed him. One such lover, Marita Lorenz, says Castro was aware of her assassination plan and offered her his gun in lieu of poison: "He kind of smiled and chewed on his cigar. I felt deflated. He was so sure of me. He just grabbed me. We made love."[193]

As late as 2000, the CIA attempted to assassinate Castro using explosives under a podium at a public speaking event in Panama. CIA agent Luis Posada was captured and imprisoned by Cuban authorities.

In addition, two hitmen on the FBI's Top Ten Most Wanted List were at one point hired by the CIA to assassinate Castro. The CIA wanted to recruit and utilize mafia hitmen because they offered cover. If the CIA was accused of an assassination, CIA officials could blame the mafia and avoid accountability.[194]

The Northwoods Memo and CIA Operation Zapata

After CIA-funded Cuban rebels failed to overthrow Castro, U.S. President John F. Kennedy (JFK) was pitched with several false flag operations. A successful false flag would help the U.S. 1% persuade the U.S. 99% that invading Cuba was justified. As described in top secret U.S. military documents, the Joint Chiefs of Staff listed a number of pretexts they could use to invade Cuba. Among them were the following:

- A "Remember the Maine" incident could be arranged in several forms.... We could blow up a U.S. ship in Guantanamo Bay and blame Cuba....
- We could develop a Communist Cuban terror campaign in the Miami area, in other Florida cities and even in Washington....
- It is possible to create an incident which will demonstrate convincingly that a Cuban aircraft has attacked and shot down a chartered

> civil airliner en route from the U.S. to Jamaica, Guatemala, Panama or Venezuela....[195]

Instead of using the military, JFK allowed the CIA to attempt yet another covert operation. CIA Operation Zapata involved training more than 1,000 Miami-Cuban exiles known as Brigade 2506. If the operation succeeded, Cuban exile José Miró Cardona would be installed as the new president of Cuba. Operation Zapata failed. Brigade 2506 was captured by Castro's army when it invaded Cuba's Bay of Pigs.[196, 197] Instead of being executed, the captured exiles were provided a televised trial. Communications technology empowered the entire world with accountability information. On that day, Cuba was showing the world how democracy works.[198, 199]

The Secret CIA Overthrow of the U.S.

Through Operation Mockingbird, agreements could be made with mass media not to propagate information critical of the CIA or its allies. Philip Graham, co-owner of the *Washington Post*, had an agreement with CIA Director Dulles to censor information in exchange for the *Washington Post* becoming the dominant newspaper in Washington D.C., which it did.[200] The *Washington Post* model ultimately became a primary product offering for the CIA: customers secretly do whatever the CIA asks of them in exchange for the CIA covertly eliminating the customer's competition.[201, 202]

Much as the CIA was seizing global communication power, it was also seizing U.S. legislative decision power. According to one *National Enquirer* article (which was never published due to CIA intervention), film director Howard Hughes had secretly taken money from the CIA and used it to provide campaign donations to U.S. politicians. Some of those politicians controlled the fate of the CIA budget through congressional committees. Each election cycle, congressional committees redirected more and more tax dollars to the CIA.[203, 204]

According to CIA Agent Robert David Steele, "The sad thing is that the CIA is very able to manipulate [the media] and it has financial arrangements with media, with Congress, with all others."[205]

CIA Agent Gilbert Greenway was even more direct in his recollection: "I remember once meeting with [CIA Deputy Director] Wisner and the [CIA] comptroller. 'My God,' I said, 'how can we spend that?' There were no limits, and nobody had to account for it. It was amazing."[206]

CIA officials could secretly buy more tax dollars for themselves. For communication purposes, a feedback loop was created. From the perspective of unwitting U.S. taxpayers and voters, campaign donations were a negative feedback loop. From the perspective of politicians and CIA officials, campaign donations were a positive feedback loop. CIA officials were gradually overthrowing the U.S. government.[207, 208]

PART II

OPTION POWER

CHAPTER 7

Controlling Economies

If You Control the Options, You Control the Decisions

Option power is an abstract concept. It can be understood as the power of having resources at your disposal. Economies are a function of resource exchanges. Economics is therefore inseparable from governance. The following are examples of resources:

- *Natural resources:* food, water, land, horsepower, oil, electricity, nuclear power, other energy
- *Human resources:* labor, sexuality, spirituality, creativity, persuasiveness, humor
- *Financial resources:* currency, loans, securities, taxes, other crowdfunding, precious metals

- *Information resources:* voting results, legislated rights, judicial decisions, publications

If you can control a group's resources (options), you can control its decisions. This principle holds true for two key categories of resources: commercial (products) and political (candidates).

Defining the Group

As discussed in the Introduction of this book, the four powers equation shows us there are only two types of government systems: democratic and nondemocratic. In a nondemocratic government system, these four powers are typically centralized in one member or less than 1% of the group. In a democratic government system, the four powers are decentralized to up to 100% of group members. To understand power, the group must be defined.

In the case of an individual, 100% of the power is vested in one person. In the case of a group of two, 50% of the power could be democratically vested in each group member. In the case of a group of three, 33% of the power could be democratically vested in each group member, and so on.

In the case of a nation, the group is defined as its citizens. Citizens are bound by the laws and other decisions made as a nation. Laws are one mechanism for centralizing and decentralizing power. Immigration and citizenship mechanisms are typically legislated to define who is a legitimate group member of a given nation. Regardless of these laws, noncitizens can always influence the power structure of a nation by exchanging resources with bona fide citizens of that nation.

Hypothetically, noncitizens, extraterrestrials, or other nongroup members can impact the power structure of a nation. For example, extraterrestrials can exchange information resources, such as advanced technology, for land resources, such as covert underground bases. Those technology resources, such as supercomputers or planet-killing weapons, could then be capitalized by the recipient nation.

Advanced technology would likely be kept secret, as it would give the recipient nation a massive advantage over other nations. To prevent other nations from capitalizing on similar exchanges, that relationship between the recipient nation and the extraterrestrials would be top secret. New social mechanisms, such as secrecy laws, would be required to classify and protect information considered to be top secret.

CHAPTER 8

Manipulating Options

Monopolies and Duopolies

Most of Earth's resources are undemocratically controlled and manipulated by monopolies (one option) and duopolies (two options). To the untrained eye, a supermarket offers an infinite-option system. For example, in the U.S., when you enter a supermarket to purchase cereal, you appear to have more than fifty options. In reality, the system is typically two options because two corporations are supplying all fifty cereals.

Corporations are social mechanisms for scaling the exchange of resources. Cereal corporations, like Kellogg Company and General Mills, offer numerous product lines, such as Frosted Flakes and Cheerios, respectively.[209] Shoppers feel empowered by the illusion of having infinite options at their disposal. In that situation, the supermarket has a monopoly while Kellogg Company and General Mills have a duopoly.

Political Party Systems

Political party systems are similar to supermarket systems. For example, the U.S. is a duopoly. U.S. voters generally have the same two options every election cycle: Democrat or Republican. Each party is its own brand, propagating polar ideologies. Every election, each party decides its own candidates. Subsequently, the wealthy can guarantee their return on investment by making campaign donations to both candidates. By funding both candidates, the campaign donor always wins.

According to one study, corporate campaign donors make $760 for every $1 they donate.[210] Campaign donors have a monopoly, and political parties have a duopoly. Like supermarket shoppers, voters feel empowered by the illusion of deciding who runs their country. Election rigging can occur by manipulating a variety of factors such as the following:

- candidacy requirements
- which candidates get on the ballot
- campaign donations
- which group members vote
- the polling stations
- the vote collection process
- the vote counting process

Political elections are the path to decision power over any group's pooled resources. They are the Achilles' heel of any democracy.

Corporate Antitrust

Election rigging is clearly undemocratic. Commercial election rigging can occur when all product options are owned by one group member. Regardless of which product each group member decides on, the group member who controls all the options always wins. Stated differently, decision power becomes monopolized. The Amazon marketplace, for example, is rapidly becoming a monopoly.

Commercial election rigging effectively occurs through antitrust. When group members who control options allow free market conditions to dictate prices, it builds trust. When these trusted group members then conspire to fix prices or otherwise covertly rig product options, such as oil corporations secretly agreeing to set all gas prices at the same level, it creates distrust. Globally, social mechanisms, in the form of antitrust laws, have been created to prevent this.

Political election rigging can occur through assassination or ballot manipulation, among other methods. A candidate cannot be elected or reelected if he or she is dead or otherwise no longer on the ballot. Examples of political election rigging are provided in the next chapter.

CHAPTER 9

Political Election Rigging

1960: The Assassination of Iraqi Prime Minister Abd Al-Karim Qasim

In 1960 the London-based Iraqi Petroleum Company (IPC) had a monopoly on oil production and exports in Iraq. After two world wars, the demand for oil skyrocketed. Almost every tank, aircraft, and naval ship on the planet required oil for propulsion. Thirty years had passed since IPC discovered oil in Iraq, and the Iraqi government wanted more than the small royalty it was paid under the IPC contract. As such, Iraq's government wanted a 20% ownership interest in Britain's IPC, as well as 55% of any profits.[211]

After his demands were ignored, Iraqi Prime Minister Abd Al-Karim Qasim reached out to other Middle East countries and created the Organization of Petroleum Exporting Countries (OPEC). OPEC wanted foreign oil companies out of the Middle East.

The CIA sprang into action, covertly looking to replace Prime Minister Qasim with a puppet leader. They found willing partners in the Iraqi military. General Ahmed Hassan al-Bakr and his cousin Saddam Hussein arranged for an armed overthrow and the arrest of Prime Minister Qasim. After a sham trial, Qasim was shot and killed. The process was efficient, and the CIA was glowing in its success.[212] According to CIA agent James Critchfield, "We really had the t's crossed on what was happening.... We regarded it as a great victory."[213]

The new partnership between the CIA and U.S. corporations was in full swing. When U.S. corporations were unable to diplomatically seize control of another country's resources, the CIA would covertly replace that country's top decision-makers. In the case of Iraq, President Al-Bakr was later succeeded by a similarly corruptible Hussein, whose security forces assassinated over 200,000 political enemies in an effort to stay in power. As a puppet for the CIA, President Hussein's net worth reached an estimated $2 billion.[214, 215]

Premeditated Airplane Disasters

Premeditated airplane disasters are convenient social mechanisms for covert agencies because they assassinate options (human resources) and destroy evidence of a crime. By destroying evidence, accountability power over any premeditated airplane disaster remains centralized in the perpetrators. Following Operation Zapata and the Bay of Pigs failure, the CIA continued its attempts to assassinate members of the Castro family, including Fidel Castro's brother Raul.

At one point, a Cuban airline pilot was recruited by CIA Agent William J. Murray. The pilot, José Nunez, anticipated he would be piloting a plane carrying Raul Castro, Cuba's minister of the armed forces, from Cuba to Prague. In the event that he died while assassinating Raul Castro, the pilot wanted the CIA to ensure his two sons would receive college educations. The assassination attempt was a failure, and Raul Castro later succeeded his brother as Cuba's leader.[216]

Statistically, an unusually high number of politicians have died in airplane disasters. According to NSA whistleblower John Perkins, the CIA has utilized airplane disasters as cover to assassinate several politicians, including presidents Jaime Roldós Aguilera of Ecuador and Omar Torrijos of Panama. In the cases of Roldós and Torrijos, the assassinations occurred two months apart. Both Roldós and Torrijos were opposed to working with U.S. corporations on multibillion-dollar projects, including oil drilling in South America and the building of the Panama Canal.[217, 218]

According to former Malaysian Prime Minister Mahathir Mohamad, the CIA was also involved in the disappearance of Malaysian Airlines Flight 370.[219]

1961: The Assassination of Congo Prime Minister Patrice Lumumba

In 1961 the CIA used U.S. taxpayer funding to undemocratically install General Joseph Mobutu Sese Seko as dictator of Zaire (now known as the Democratic

Republic of the Congo). Under Operation Barracuda, the CIA paid General Mobutu $250,000 to assassinate Prime Minister Patrice Lumumba, the first democratically elected leader of Zaire. Lumumba was executed by firing squad.[220, 221]

Over the next three decades, the CIA used the Congo as a staging point for covert actions throughout Africa. Just like they were doing in the Middle East, South America, Central America, Europe, and Southeast Asia, the CIA was covertly installing puppet leaders in every African country.[222, 223, 224]

Because the CIA controlled almost all U.S. media, the American people were cut off from accountability information involving CIA-created havoc. Twenty-something-year-old CIA agents, possessing fraudulent identifications, bags full of U.S. tax dollars, weapons, and a wealth of training were embedded in almost every country on the planet.

The Assassination of Dominican President Rafael Trujillo

In 1961, the CIA used U.S. taxpayer funding to undemocratically install Juan Bosch Gaviño as president of the Dominican Republic. Then–Dominican President Rafael Trujillo was shot to death by assassins using CIA-supplied semiautomatic M1 carbines. At the time, Trujillo owned more than 100 companies, including monopolies over Dominican sugar, salt, tobacco, meat, and dairy. The assassination of Trujillo enabled instant CIA control of the Dominican Republic, as well as neighboring countries such as Haiti.[225, 226]

Understanding he might be assassinated next, Haitian President François "Papa Doc" Duvalier began secretly cooperating with the CIA. Papa Doc proposed a referendum to the Haitian constitution that made him "president for life" and ensured that his son, Jean-Claude "Baby Doc" Duvalier, would automatically be next in line. The referendum passed with Papa Doc receiving an improbable 99.9% of the vote.[227, 228]

Papa Doc proved to be a madman. After uncovering an imminent coup attempt by the KGB-backed head of Haitian security forces, Clément Barbot, Papa Doc ordered a manhunt. Upon receiving information that Barbot had somehow transformed into a black dog, Papa Doc ordered all black dogs executed.[229] This was the man the CIA had decided to align forces with. Baby Doc, who took power in 1971 after his father died, was ruthless as well. After twenty-nine years of poverty and over 30,000 assassinated political enemies, the CIA helped Baby Doc flee a violent Haitian revolt in 1986.[230, 231]

1962: The Assassination of Vietnamese President Ngo Dinh Diem

In 1962, the CIA used U.S. taxpayer funding to undemocratically install General Duong Van Minh as president of Vietnam. After assassins received cash payments from CIA Agent Lucien Conein, Vietnamese President Ngo Dinh Diem was forced into an M113 tank, stabbed repeatedly with a bayonet, and then shot in the back of his head with a revolver. The body was dumped in an unmarked grave at the home of the U.S. Ambassador to Vietnam Henry Cabot Lodge.[232, 233, 234]

1963: The Assassination of U.S. President John F. Kennedy

Following the Bay of Pigs debacle, U.S. President Kennedy believed he had been deliberately set up for political failure by the CIA. As a result, he fired CIA Director Allen Dulles and told an aide he wanted to "splinter the CIA into a thousand pieces and scatter it to the winds."[235]

In November 1963, JFK was assassinated. According to the FBI, the lone assassin was Lee Harvey Oswald, although the facts surrounding the assassination are still in question. Forensic evidence indicated that multiple bullets entered both the front and back of JFK's body, implying multiple shooters. Upon arrest, Oswald insisted he did not kill JFK and claimed he was a patsy. Two days after JFK's assassination, Oswald himself was assassinated.[236, 237]

Multiple whistleblowers have linked Oswald to the CIA. At a closed session of the U.S. Congress in 1978, CIA official James Wilcott disclosed that Oswald was a paid operative. Morita Lorenz also linked the CIA to JFK's assassination. Lorenz is the mother of Andres Vazquez, the son of Fidel Castro. According to Lorenz, she and Oswald trained with the CIA for Operation 40, an effort to overthrow Castro after the failed Bay of Pigs invasion. Lorenz, Oswald, Frank Sturgis (a Watergate burglar), and other CIA assets allegedly travelled together to Dallas prior to JFK's assassination.[238, 239, 240, 241]

Under the Assassination Records Collection Act of 1992, all JFK assassination investigation documents should have been published in 2017. As of 2020, accountability power over the JFK assassination continues to be centralized. Over 10,000 documents unlawfully remain unpublished.[242]

Unelected U.S. Decision-makers

Following the assassination of JFK and pursuant to the Twenty-fifth Amendment to the U.S. Constitution, Vice President Lyndon B. Johnson (LBJ) became president without ever receiving a single vote.[243, 244, 245]

LBJ had a history of illegitimately gaining office. Two decades earlier, Democratic Party officials rigged the 1948 Texas general election, launching LBJ's political career. Years later, Texas election judge Luis Salas confessed his involvement in rigging the 1948 vote counts. According to Salas, vote counters repeatedly "provided an extra 200 votes for [LBJ] merely by changing the 7 in '765' to a 9."[246]

Law enforcement agencies, like the courts, are perfect examples of social mechanisms designed to decentralize accountability power. In the U.S., the FBI is the lead law enforcement agency. People who were proximate to LBJ have since come forward with information that incriminates both LBJ and FBI Director J. Edgar Hoover, two of the most powerful government officials in the U.S. at the time of JFK's murder.

Kompromat on FBI Director J. Edgar Hoover

CIA officials understood that the FBI was arguably the greatest threat to its existence. To ensure its survival, the CIA was installing its agents inside the FBI. Ideally, they would want a covert CIA agent, like L. Patrick Gray, to replace Hoover. At some point, the CIA acquired kompromat, or damaging information, in the form of controversial photos that gave the CIA leverage over Hoover.[247]

According to statements from Madeleine Duncan Brown (LBJ's former girlfriend and mother of his son), Hoover, LBJ, and CIA officials held a secret meeting in Dallas the night before JFK was assassinated.[248] Former CIA Agent E. Howard Hunt has also stated that LBJ, CIA officials, and FBI officials conspired to assassinate JFK, specifically calling out CIA Agents Bill Harvey and David Morales.[249, 250] Another known CIA asset, Antonio Veciana, has placed CIA Agent David Atlee Phillips in Dallas with Oswald.[251]

Heightened scrutiny is necessary when information incriminates the people empowered with accountability over the crime in question. In this case, JFK's assassins were supposed to be held accountable by Hoover, the lead investigator. Two days after JFK's assassination, however, Hoover documented the following: "The thing I am concerned about, and so is [the U.S. deputy attorney general], is having something issued so we can convince the public that Oswald is the real assassin."[252]

At a minimum, the memo implies the willingness of the FBI director to propagate an investigative conclusion after only two days of investigating.[253] At worst, it suggests Hoover was an accomplice to JFK's murder. At risk of incriminating its founding director and undermining its own credibility, the FBI's consistent decision not to provide follow-up investigations has been a self-serving abuse of power.[254]

1964: The Assassination of Greek Prime Minister Georgios Papandreou

In 1964 the CIA used U.S. taxpayer funding to undemocratically force Greece's Prime Minister Georgios Papandreou out of office. Greek Colonel George Papadopoulos, a CIA asset, was eventually installed as prime minister. After six years, Papadopoulos was arrested for high treason and sentenced to life in prison. LBJ had a revealing response:

> "Fuck your parliament and your constitution. America is an elephant. Cyprus is a flea. Greece is a flea. If these two fleas continue itching the elephant, they may just get whacked good... We pay a lot of good American dollars to the Greeks, Mr. Ambassador. If your Prime Minister gives me talk about democracy, parliament and constitution, he, his parliament and his constitution may not last very long."[255]

The CIA also considered the politically active son of Georgios Papandreou, Andreas, to be a political threat. CIA agent Gustav Avrakotos directed Greek military officials to "Shoot the motherfucker—because he's going to come back to haunt you."[256]

Andreas Papandreou fled to Sweden with his wife and children in 1967. Georgios Papandreou died under house arrest in 1968. As predicted, Andreas

Papandreou returned to haunt the CIA after he was elected prime minister of Greece in both 1981 and 1993. Greece provided the CIA with a valuable lesson: When given the chance, murder the whole family. Otherwise, charismatic sons like Andreas Papandreou or JFK Jr. may come back to haunt you. JFK's son, John F. Kennedy Jr., died in an airplane disaster in 1999.[257]

CHAPTER 10

White Power

1965: The Assassination of Malcolm Little

Malcolm Little was born in Nebraska in 1925. His surname was presumably the labeling of a slave trader who thought Malcolm's ancestors were physically small. When Malcolm was six, his father was fatally hit by a car driven by a member of the Black Legion. The Black Legion was a Midwest white supremacist group that hunted black people for sport. According to Malcolm, his father and four uncles were all killed by white supremacists. After being disempowered of family and home, he moved from foster family to foster family and was eventually sentenced to ten years in prison for burglary.[258]

According to Malcolm, prison did not feel like prison. He said he felt truly free. Empowered with a sense of community, Malcolm joined the Nation of Islam (NOI), a black supremacist group, and changed his name to Malcolm X. The "X" served as a placeholder for his unknown African surname.[259] After parole, Malcolm X took up a leadership role within NOI New York.

After two New York Police Department (NYPD) officers brain-damaged a man using nightsticks, more than 500 protesters gathered. Malcolm X diffused the violence with a hand gesture to fellow NOI members. All the protestors quietly left. As one NYPD officer hypocritically described it, "No one man should have that much power."[260]

By 1965 the NYPD and FBI had Malcolm X under surveillance and were assigning numerous undercover agents to the NOI. After a United Nations general assembly in New York and a meeting with Cuba's Fidel Castro, Malcolm X was on the CIA's radar as well. The CIA had already covertly subdued NOI leader Elijah Muhammed, who was taped having extramarital sexual affairs with NOI staffers.[261] After an inspiring trip to Saudi Arabia as a guest of Prince Faisal, Malcolm X denounced the NOI and publicly endorsed nonviolent Sunni Islam.[262]

Malcolm X was exhibiting a powerful influence over a global community. He converted world-champion boxer Cassius Clay to Islam, later resulting in Clay's name change to Muhammed Ali. Several months later, Malcolm X was preparing to speak at a rally when one of the attendees yelled, "Nigger!"[263] In the ensuing pandemonium, Malcolm X was fatally shot in the chest with a sawed-off shotgun.[264]

1966: The Ku Klux Klan

For centuries, governing power on Earth has been primarily centralized in white males—a good ol' boy network. At a high level, the foundation of white power has been an effort to keep nonwhites out of positions of power, where they could take control of resources. In the wake of the American Civil War, a terrorist group emerged called the Ku Klux Klan (KKK). Referred to as the Invisible Empire by some members, the KKK was invigorated by a 1915 propaganda film, *The Birth of a Nation.*[265]

Having little to no public accountability, KKK operations allowed whites to maintain centralized power. As long as political offices, law enforcement agencies, and businesses were run by whites, power inequality in favor of whites would persist. Whites would continue hiring and voting for whites. The KKK rigged political elections in favor of whites and assassinated popular nonwhite candidates and civic leaders.

When they assembled, Klan members would hide under pointy white hoods, refer to their leaders as wizards, and plan covert operations. KKK wizards would mastermind covert operations that, as if by magic, ensured whites maintained centralized power. On one occasion in 1966, an Alabama KKK wizard decided that a group of Klansmen would go out and "grab a negro and scare [the] hell out of him."[266] The would-be victim was Edward Aaron. As they sliced off Aaron's testicles, one Klansmen shouted, "You think nigger kids should go to school with my kids?"

Aaron survived the attack primarily because the Klansmen poured turpentine on his genitals to intensify the pain. Inadvertently, the turpentine prevented the wound from becoming infected, effectively saving Aaron's life. After their arrest, two of the Klansmen turned state's witnesses, testifying against the other four Klansmen in exchange for reduced sentences. The remaining four Klansmen were convicted in a public jury trial. Justice appeared served.[267]

1967: Pardoning Bad Behavior

The KKK had friends in high places. All four Klansmen convicted of attacking Edward Aaron were set free following a pardon from Alabama governor and future Democratic U.S. presidential candidate George Wallace. In retaliation for betraying the KKK, the two state's witnesses were left to serve their full reduced sentences.

Wallace's decision to pardon was an abuse of power. Wallace had a clear conflict of interest. The decision to pardon the four Klansmen served the personal interests of Wallace, not the public interest of accountability, as evidenced by the jury verdict.[268] KKK leadership understood they could always maintain this option power if "their people" perpetually held positions vested with decision power, such as the office of the U.S. president.

1968: The Assassination of U.S. Presidential Candidate Martin Luther King Jr.

Michael King Jr. was born in Atlanta, Georgia, in 1929. After his father, a college educated pastor, visited Lutheran Reformation sites in Europe, Michael was renamed Martin Luther King Jr. (MLK). When MLK was six, his best friend, who was white, left for a white-only school and was prohibited from associating with the King family and other nonwhites. According to MLK, he was then "determined to hate every white person."[269]

At the age of fifteen, MLK was admitted to Morehouse College. On a trip to Washington, D.C., MLK noted, "The white people here are very nice." MLK was suddenly empowered with the freedom to decide his own actions, noting, "We go to any place we want to and sit anywhere we want to."[270] He learned that the U.S. Constitution is a social agreement between all the American people, regardless of skin color. MLK understood that every member of a group must be equally empowered to maintain freedom and justice. He famously described it in the following words:

> "I have a dream that one day this nation will rise up and live out the true meaning of its creed: 'We hold these truths to be self-evident, that all men are created equal.' I have a dream that one day on the red hills of Georgia, the sons of former slaves and the sons of former slave owners will be able to sit down together at the table of brotherhood. I have a dream that one day even the state of Mississippi, a state sweltering with the heat of injustice, sweltering with the heat of oppression, will be transformed into an oasis of freedom and justice. I have a dream that my four little children will one day live in a nation where they will not be judged by the color of their skin but by the content of their character."[271]

As well as almost any American ever had, MLK asserted his communication powers through religion, speech, press, and assembly. At some point, MLK communicated a desire to run for U.S. president.[272] This did not go over well with many U.S. government officials, who likely overheard the conversation using warrantless surveillance techniques. Under Operations COINTELPRO, MHCHAOS, and LANTERN SPIKE, the FBI, CIA, and U.S. military, respectively, were conducting warrantless domestic surveillance on MLK. No search warrant could be obtained because MLK had not committed a crime. There was no legal justification for government surveillance.[273, 274, 275]

The surveillance conducted by all three agencies was unequivocally unconstitutional and illegal. Every conversation, sexual encounter, and trip to the bathroom was illegally recorded, in violation of the Fourth Amendment to the U.S. Constitution. White law enforcement officials did not like a black man holding them accountable to the U.S. Constitution. Any embarrassing recording could be used by U.S. government officials to secretly blackmail MLK.[276]

In 1968 MLK was assassinated. According to the FBI, the lone assassin was James Earl Ray, but the facts surrounding the assassination are still in question. Until his death in 1998, Ray insisted he did not kill MLK. During a civil lawsuit filed by MLK's family in 1998, a jury concluded that business owner Loyd Jowers, Memphis police officer Earl Clark, and other unidentified government agents had assassinated MLK and covered up the crime. Following the civil trial, which provided new evidence, the FBI again failed to investigate.[277]

The Assassination of U.S. Presidential Candidate Robert Kennedy

Three years after the assassination of his brother JFK, Senator Robert Kennedy gave the following speech:

> "Hand in hand with freedom of speech goes the power to be heard, to share in the decisions of government which shape men's lives. Everything that makes man's life worthwhile—family, work, education, a place to rear one's children and a place to rest one's head—all this depends on decisions of government. All can be swept away by a government which does not heed the demands of its people. Therefore, the essential humanity of men can be protected and preserved only where government must answer, not just to the wealthy, not just to those of a particular religion, or a particular race, but to all its people."[278]

Two years later, Robert Kennedy was assassinated moments after winning the California presidential primary election. According to the FBI, the assassin was Sirhan Sirhan, but the facts surrounding the assassination are still in question. Until his death, Sirhan insisted he did not remember killing Robert Kennedy.[279, 280, 281]

CHAPTER 11

Double Agents

1969: The Attempted Assassination of Cambodia's King Norodom Sihanouk

In 1969 the CIA, under Operation Cherry, attempted to assassinate King Norodom Sihanouk and install Prime Minister Lon Nol as the leader of Cambodia. CIA agents were given fraudulent identifications and doubled as U.S. military intelligence officers. A local rebel, Inchen Lam, assisted the CIA but ended up being a KGB double agent. Lam was responsible for sabotaging the parachutes of CIA agents sent to assassinate Sihanouk.[282, 283]

Following a CIA-rigged election, Nol became president of Cambodia. In response, Sihanouk fled to China and used Beijing Radio to orchestrate protests against Nol. Armed with CIA weapons and intelligence, Nol violently suppressed uprisings as they arose across the country. According to an eyewitness in one village, "Tanks flattened the people, two or three hundred died. I rode my bicycle there and saw all the bodies. Some people were not yet dead, and were lying there, waving flags and their hands, saying 'long live [Sihanouk]'. Then soldiers brought tanks and flattened them all [again]."[284]

Since the CIA-driven destabilization of the 1970s, Cambodia has succumbed to poverty and human trafficking. Children are treated like financial resources and sold into slave labor and sex trades.[285]

1970: The Assassination of Chilean Commander-in-Chief René Schneider

In 1970 the CIA and KGB were fighting for control of Chile through the presidential election process. While the KGB was secretly funding one candidate, the CIA and U.S. corporations were secretly funding the other candidate.

After the KGB's candidate won the election, the Chilean military's commander-in-chief, René Schneider, refused to assist the CIA with an overthrow. According to Schneider, "The armed forces are not a road to political power nor an alternative to that power. They exist to guarantee the regular work of the political system, and the use of force for any other purpose than its defense constitutes high treason."[286]

In response to Schneider's blowing the whistle on the CIA's plans, a double agent sledgehammered a window on Schneider's vehicle and shot him to death.

1971: The Political Assassination of Australian Prime Minister Gough Whitlam

In 1971 Gough Whitlam rose to power in the Australian Labor Party and was elected prime minister. Whitlam instituted free education and healthcare for all. Whitlam also wanted Australia to retake ownership of its oil and mining properties controlled by U.S. corporations.

Whitlam had discovered that Australia's Pine Gap research facility was secretly funded by the CIA. Run by CIA Agent Richard Stallings, Pine Gap was used to spy on Russia, find Australian gold deposits for U.S. mining corporations, and coordinate U.S. bombing strikes in Vietnam.[287, 288]

A secret CIA-coordinated coup by Australian Governor-General John Kerr ensured Whitlam's forced removal from office. Kerr had been a member of an Australian intelligence unit during World War II and was receiving payments from the CIA through a San Francisco–based shell corporation called the Asia Foundation.[289, 290]

1972: The Assassination of Dorothy Hunt (United Airlines Flight 553)

In 1972 Dorothy Hunt, the wife of CIA agent E. Howard Hunt, died when United Airlines Flight 553, travelling from Washington, D.C., to Chicago, crashed in an Illinois suburb, killing forty-five people. As discovered after the crash, Dorothy Hunt had boarded the plane carrying $10,000 in cash.[291] Her husband had been indicted three months earlier for the Watergate burglary, and Dorothy Hunt was blackmailing CIA officials for money.[292]

According to President Nixon's special counsel, Charles Colson, "Nobody controls the CIA. I mean nobody. If the CIA really has infiltrated this country to the extent I think it has, we ain't got a country left."[293]

Colson believed the CIA was responsible for the crash of Flight 553 as a means to stop Dorothy Hunt. According to CIA agent James McCord, Dorothy Hunt was supplying the Watergate burglars with cash for their legal expenses.[294]

1973: The Assassination of Chilean President Salvador Allende

In 1973 the CIA-backed Chilean military surrounded the presidential palace of Chile's new president, Salvador Allende. According to Chilean military officials, Allende committed suicide using an AK-47. The gun had been a gift from Fidel Castro. According to the autopsy report, "Allende died of two shots fired from an assault rifle that was held between his legs and under his chin and was set to fire automatically. The bullets blew out the top of his head and killed him instantly."[295]

The CIA decided Chilean Army Chief of Staff Augusto Pinochet would be installed as the next president of Chile. No election was held. Pinochet immediately banned labor unions and privatized state-owned resources. Over 2,000 protestors were killed, and more than 20,000 imprisoned. Chileans who protested against the CIA-backed Pinochet were tortured and drowned to death using submersion in buckets of urine and excrement.[296, 297, 298, 299]

1974: The Assassination of Jamaican Political Activist Bob Marley

In 1974 the CIA began providing weapons to Jamaican Labor Party (JLP) operatives in an attempt to overthrow Prime Minister Michael Manley.[300] Manley was undermining the U.S. by, among other things, supporting neighboring Cuba. In response, the CIA provided Lester Coke with the guns and money needed to advance the interests of JLP politicians, including Edward Seaga.[301]

According to CIA agent Philip Agee, "The CIA was using the JLP as its instrument in the campaign against the Michael Manley government. I'd say most of the violence was coming from the JLP, and behind them was the CIA in terms of getting weapons in and getting money in."[302]

Central to the Manley 1976 reelection effort was musician and activist Bob Marley. In 1976, Coke's men sprayed Marley's home with gunfire.[303] Rita Marley, the musician's wife, was shot in the head.[304] Despite being shot in the arm, Marley gave a concert a few days later and endorsed reelecting Manley at the event. There, a double agent named Carl Colby, the son of CIA Director William Colby, posed as an admiring member of a film crew and presented Marley with a gift: a deadly pair of boots boobytrapped with a carcinogen-ladened wire inserted in the toe. Within months, Marley's toe was amputated. Meanwhile, cancer spread throughout his body.[305]

The CIA eventually took control of Jamaica. During the 1980 Jamaican elections, the CIA delivered to the JLP an improbable fifty-one out of sixty congressional seats, and Seaga was installed as prime minister.[306]

1975: The Assassination of Saudi Arabia's King Faisal

In 1973, Saudi Arabia's King Faisal spearheaded the OPEC oil crisis by cutting off the U.S. and other countries from Middle East oil. To restore oil supplies, King Faisal demanded that the U.S. stop supplying weapons to Israel. This leveraging did not go over well with CIA officials.[307]

In 1975 King Faisal's nephew returned to Saudi Arabia and made an unexpected appearance at the royal palace. The nephew had been attending college in California and Colorado, where he allegedly became addicted to LSD. When the king saw his nephew, he tipped his head, expecting to receive a monarchal kiss. Instead, Faisal's nephew pulled out a gun and shot his uncle twice in the head.[308]

King Faisal's brother Khalid was immediately appointed king by a group within the royal family. Khalid, who was understandably friendlier towards the U.S. and Israel than his dead brother, subsequently made more than $2 billion by skimming oil revenue.[309]

CHAPTER 12

Intimidation

1976: The Assassination of Brazil's President João Goulart

João Goulart was a rancher, attorney, and Brazil's minister of labor when he publicly stated he was opposed to "parasitic, speculative, exorbitant and short-sighted capitalism in profit."[310] Goulart was eventually elected president of Brazil in 1961. At that point, the CIA stepped in. Lincoln Gordon was the U.S. ambassador to Brazil. As Gordon described it, the CIA was prepared to "intervene militarily to prevent a leftist takeover."[311]

The CIA was undemocratically overthrowing another democratically elected leader simply because he was a leftist. There was no evidentiary basis for removing President Goulart. He had committed no crime. He simply opposed short-sighted capitalism. CIA resources would be delivered to Brazilian military officials by the U.S. Navy using an "unmarked submarine to be off-loaded at night in isolated shore spots in the state of São Paulo south of Santos."[312]

To intimidate the population, CIA assets ordered Brazilian military troops into the streets. To avoid a bloodbath, Goulart fled with his wife and kids to Uruguay and eventually Argentina. In 1976 as Goulart was plotting his return to Brazilian politics, he was poisoned to death.[313]

The Assassination of Native American Activist Anna Mae Aquash

Anna Mae Aquash was an influential Native American activist and member of the American Indian Movement (AIM), a group committed to equality and protecting natural resources. She participated in the 1972 Trail of Broken Treaties March on Washington, meant to bring attention to Native American inequality issues. Aquash also participated in the occupations of Wounded Knee and Anishinaabe Park. She was responsible for promoting drug-free living and for recruiting Hollywood celebrities to support AIM. Aquash was a close friend of AIM member Leonard Peltier, a Native American activist wanted for killing two FBI agents.[314]

According to witnesses, one of the agents hunting for Peltier, FBI Agent David Price, told Aquash, "If she didn't start cooperating that he would see that she died."[315] Aquash was not intimidated and did not cooperate. That winter, a body was found in a ravine on a reservation in North Dakota. Agent Price declared the body unidentifiable and had the hands cut off for fingerprinting prior to dumping the body in a pauper's grave. According to an autopsy by the U.S. Bureau of Indian Affairs (BIA), the unidentifiable person had died from exposure. FBI headquarters then announced they had found the body of Anna Aquash and informed Aquash's family she had "died from natural causes."[316]

Accountability requires evidence. Having centralized accountability power, the FBI, BIA, and CIA were able to conceal evidence, or so they thought. After her family exhumed the body through a court order, a second autopsy revealed a bullet in the back of Aquash's head—a small oversight by FBI and BIA experts. Aquash had been executed from behind.[317]

1977: The Assassination of Argentina's Mothers of the Plaza de Mayo

After the death of Argentina's President Juan Perón, his wife, Vice President Isabel Perón, ascended to the presidency. She was the first female to hold the title of president in the modern world. Naturally, the CIA stepped in, covertly providing weapon resources to corrupt Argentine military officials.[318] During the overthrow, Isabel Perón was arrested. Argentina Army Commander Jorge Videla was installed as president. The Videla regime was responsible for the death or disappearance of roughly 30,000 political dissidents.[319, 320]

Forensic and testimonial evidence has since been used to hold the Videla regime accountable. In the wake of the coup, Videla officials began sex trafficking local children to Videla supporters. Mothers of the missing children began protesting as a group called Mothers of the Plaza de Mayo. The mothers brought unwanted international attention to the Videla regime, exposing human rights violations in Argentina. In 1977 the unidentifiable remains of three women subsequently washed up on a local beach.[321, 322]

In 2005 anthropologists identified the remains as those of the three missing women who founded the Mothers of the Plaza de Mayo. According to whistleblowers and investigators, the women were victims of "death flights." The Argentine army had kidnapped the three women, flown them above international waters, and pushed them out of a plane.[323, 324]

1978: The Assassination of Afghanistan's President Mohammad Daoud Khan

In 1978 Afghan military officials seized government radio towers as well as the presidential palace. President Mohammad Daoud Khan and his family were executed. President Khan was replaced by Nur Muhammad Taraki. Taraki and Hafizullah Amin had orchestrated the overthrow and considered each other brothers in revolution.[325] The following year, backed by the CIA, Amin had Taraki executed and was installed as president.[326, 327]

Three months later, the KGB assassinated Amin, and Russia invaded Afghanistan. Fearing Russia intended to invade additional Middle East oil nations, the CIA launched Operation Cyclone, secretly providing funding and weapons to "freedom fighters" like Osama Bin Laden.[328, 329]

1979: The Assassination of Pakistan's Zulfiqar Ali Bhutto

In the late 1970s the CIA covertly decided the next leader of Pakistan. With the backing of the CIA, Pakistani Army General Muhammad Zia-ul-Haq overthrew Prime Minister Zulfiqar Ali Bhutto and declared martial law. Zia-ul-Haq immediately adopted pro-U.S. policies, including providing military resistance against Russia. Zia-ul-Haq was instrumental in helping the CIA orchestrate Operation Cyclone. After two years in custody, Bhutto was ultimately executed in 1979.[330, 331]

The Political Assassination of El Salvador's President Carlos Romero

In 1979 Salvadoran military officials arrested and exiled President Carlos Romero. Although Romero was a CIA asset, his human rights violations had become so intolerable, even to the CIA, that the decision was made to replace him. Romero's overthrow ignited a twelve-year civil war, resulting in approximately 16,000 civilian murders by the new regime.[332]

During protests at Rio Sumpul, approximately 600 civilians were killed, mostly women and children. According to witnesses, the CIA-trained soldiers threw babies into the air and slashed them to death with machetes.[333]

During protests at Chalatenango, approximately 450 civilians were killed. According to witnesses, the CIA-equipped Salvadoran national guard executed four American churchwomen.[334, 335]

During protests at El Mozote, approximately 800 civilians were killed. According to witnesses, members of the CIA-trained Atlacatl Battalion raped the youngest girls, slit their throats, and hung them from trees. Everyone in the town was killed.[336]

Like Afghanistan, these incidents help demonstrate how foreign governments are not pieces on a chess board that can be immaculately positioned and repositioned. Social power structures are highly unstable.

1980: The Attempted Assassination of U.S. President Ronald Reagan

In 1976 former CIA Agent George H.W. Bush was appointed CIA director. Four years later, he lost the Republican presidential primary to Ronald Reagan. Republican Party officials then decided to add Bush to the Republican ticket as vice president. After Reagan was elected president, the former CIA director was only one bullet away from becoming the U.S. president.[337]

Two months after Reagan took office, an assassination attempt was made. According to the FBI, the assassin was a clinically ill John Hinckley, but the facts surrounding the assassination attempt are still in question. On that night, Scott Hinckley, John's brother, was scheduled to have dinner with Neil Bush, the vice president's son. John's father, a Texas oil millionaire, was a substantial donor to the Bush campaign.[338, 339]

Despite motive, FBI Director William Webster never investigated the Hinckley family. Any reasonable investigator would have followed up those leads. The Hinckley leads either were not pursued or were covered up. FBI Director Webster was subsequently promoted to CIA director.[340] One could reasonably presume, based on historical precedence, that Webster buried the leads, quid pro quo, in exchange for the highest-ranking position at the CIA.

Creating a Global 1%

According to one study, the CIA has manipulated over eighty international elections. This total only represents *known* election meddling. Most recently, the U.S. 1% helped overthrow Honduras. After the Honduran people democratically elected President Manuel Zelaya, he raised the minimum wage and resisted privatization by U.S. corporations. Zelaya was subsequently arrested and deported by the Honduran military. The leader of the overthrow was Romeo Vásquez Velásquez, a Honduran general previously trained at the U.S. Army's School of the Americas.[341, 342]

By the 1980s the CIA had a proficiency for motivating government officials to overthrow their own governments. If politicians could accommodate the CIA's covert agenda, they could suddenly ascend into positions of power. The CIA would provide weapons and funding as well as train a security team to protect a puppet leader. Security teams would find, torture, and kill political enemies, enabling the puppet leader to remain in power.[343]

Training those security teams often took place at the U.S. Army's School of the Americas in Panama. A

1992 Pentagon investigation revealed that the School of the Americas taught guerrilla execution, extortion, physical abuse, and coercion.[344, 345] According to Pentagon representatives, the "School of the Americas students were taught to imprison and execute their opponents. To use motivation by fear. To subvert the press. And to use torture, blackmail, and truth serum to obtain information."[346]

U.S. Army Major Joe Blair described it thusly: "The doctrine that was taught was that if you want information, you use physical abuse, false imprisonment, threats to family members, and killing. If you can't get the information you want, if you can't get that person to shut up or stop what they're doing, you assassinate them—and you assassinate them with one of your death squads."[347]

The CIA's actions were anything but democratic. Political enemies were labelled communists, socialists, or more generally, leftists and arrested or killed by death squads even when they were nothing more than beleaguered laborers or advocates of democracy. Asserting the right to communicate or the right to accountability was fatal.

U.S. Senator Daniel Inouye described it this way in 1987: "There exists a shadowy government with its own air force, its own navy, its own fundraising mechanism, and the ability to pursue its own ideas of the national interest, free from all checks and balances, and free from the law itself."[348]

Through the CIA, option power over global resources was being centralized in a global 1%.

PART III

DECISION POWER

CHAPTER 13

Financial and Other Incentives

1981: Decisions Involving Research

Decision-making involves weighing interests, such as personal, political, business, or national security. It creates conflicts of interest. Making the right decision can be difficult. In the 1940s the U.S. government decided to create the CIA to prevent the rise of another Hitler and preferably avoid World War III.

At the height of the CIA's success, CIA officials were deciding the fate of the entire planet. A CIA agent became the U.S. president, and a CIA asset became the Russian president. Inside CIA headquarters, success stories propagated quickly while failures remained compartmentalized. CIA employees did not have the full picture. Still closely guarded secrets, the number of horrific CIA failures was beginning to add up. Each failure left its victims feeling disempowered. The more people felt disempowered, the angrier they became. The angrier they became, the more likely they were to turn violent.

Harvard University provided the CIA another valuable lesson in what causes humans to turn violent. During World War II, the CIA hired psychologist Henry Murray to create a psychological profile of Adolf Hitler. After World War II, Murray taught at Harvard and secretly provided the CIA with psychological profiles for hiring new agents.[349] CIA officials were hiring agents who would follow orders without asking questions. Accepting financial incentives from the CIA's project MK-ULTRA, Professor Murray also began conducting experiments on unsuspecting Harvard students.[350] The goal was to develop techniques for breaking down a person's ideals and belief systems.[351]

One of those individuals was a sixteen-year-old boy-genius living on scholarship in a Harvard dormitory. As part of the experiments, the boy was asked to discuss his ideals and beliefs. Experimenters then deliberately crushed those ideals and beliefs, documenting how the boy responded to the humiliation. The boy's name was Theodore Kaczynski. According to one of Kaczynski's professors, it was "not enough to say he was smart."[352] After graduating from Harvard, Kaczynski moved to the Montana wilderness and became a serial killer.[353]

Living in a cabin with no electricity or running water, Kaczynski began sending bombs by U.S. mail. The bombs, which were designed to detonate once opened, were machined and assemble in the cabin, wiped of any fingerprints, and then mailed directly to victims.[354] Kaczynski was nicknamed the "Unabomber" after the FBI case name UN-A-BOM ("UNiversity and Airline BOMber"). During his murder trial, Kaczynski was examined by the government's forensic psychiatrist. Through personal conversations, the psychiatrist linked the onset of Kaczynski's anger to the CIA experiments at Harvard.[355, 356]

1982: Decisions Involving Education

In the 1980s there was no Internet. It was more difficult to hold people accountable. Varying accounts described Rick Singer as anything from the quarterback of an intramural football team at Trinity University to a four-sport athlete at Texas A&M University.[357] Charismatic and street smart, Singer had all the tools of a good con man.

In the process of fraudulently gaining admission to several colleges and then becoming a college basketball coach, Singer had a revelation. He recognized that college admissions officials had centralized decision power. These officials had life-changing power over applicants, and the applicants had wealthy parents. Singer created a social mechanism for inducing college admissions officials to abuse their power.

Over the next thirty years, parents paid Singer upwards of $500,000 to guarantee their children's admission to schools like the University of Southern California (USC), Stanford University, and Yale University. In some cases, Singer would pay roughly $10,000 for a testing expert to take the SAT or ACT admissions exams on behalf of his applicant. In particular, Singer found a path to power by bribing decision-makers within university athletics programs. He would "donate" anywhere from $10,000 to $250,000 to cash-desperate college sports programs like tennis or rowing.

Coaches and athletic directors had conflicts of interest. They had personal interests in keeping their programs well-funded. Donor funding equated to job security. Quid pro quo, the coaches and athletic directors who benefited would then persuade their respective admissions officials to admit Singer's applicants on athletic scholarships even if they were not athletes. In one instance, the applicant's image was photoshopped over someone else's face to create the impression that the applicant actually played a particular sport. In another instance, a successful applicant for a USC rowing scholarship posed for photos with rowing equipment despite never having taken part in the activity.

About the same time Singer was identifying his cash cow, *U.S. News and World Report* began publishing annual rankings of universities. The issue was a bestseller. In the 1990s admissions officials began adapting their schools to fit the magazine's scoring criteria. It was a disaster. To the detriment of higher education, universities began spending millions of dollars to inefficiently reconfigure curriculums and improve their *U.S. News* rankings. Similar to Google search rankings, Amazon product ratings, J. D. Power product rankings, consumer safety ratings, or Moody's credit ratings, decision power was being centralized in a middleman.

1983: Decisions Involving Business

For decades, DuPont Chemical Corporation made millions of dollars selling a product called Teflon, a substance used to coat pots, pans, carpet, upholstery, and clothing. Teflon was popularized for its ability to prevent food from sticking to pots and pans during cooking. By the 1980s scientists at DuPont had discovered that perfluorooctanoic acid (PFOA), a compound used to produce Teflon, was carcinogenic.

While DuPont employees were birthing deformed babies and dying from cancer, company decision-makers looked the other way. They had conflicts of interest. Terminating the Teflon product line would have been unprofitable. Similarly, lawful disposal of chemical waste, like PFOA, would have been expensive. In response, DuPont's decision-makers secretly dumped PFOA into West Virginia lakes and streams.[358]

Over time, PFOA leached from lakes and streams into the local groundwater supply. Residents were ingesting and absorbing PFOA from their faucets and showers. When a farmer lost most of his cattle to cancerous tumors, he suspected the water supply contained chemicals. The cows had been drinking from streams covered in brown foam. The farmer filed a lawsuit against DuPont, but he and his wife died from cancer before the suit was resolved. DuPont dragged its feet for fifteen years before finally agreeing to pay over $600 million in damages to the families of the deceased.[359]

Under U.S. laws, corporations such as DuPont and Enron are self-regulating. Self-regulation centralizes accountability power. The problem is compounded when executive pay structures leverage personal financial incentives. Corporate decision-makers, such as CEOs, are rewarded with millions in cash bonuses for short-term stock price increases, regardless of long-term harm. These personal financial incentives conflict with public interests such as environmental protection, economic security, consumer safety, and employee safety.

Financial incentives for individuals create greater risk for the 100%. Imagine if hospitals began providing heart surgeons with financial incentives, such as $10,000 cash bonuses for heart transplants completed in less than eight hours. The goal would be rapid patient turnover and increased profit for the hospital. In response, some heart surgeons would likely speed up their procedures. As a result, they may skip washing their hands or rush to complete procedures, inadvertently leaving a sponge or some other device in a patient's chest cavity.

CHAPTER 14

Conflicts of Interest

1984: Decisions Involving Sex

The bondage-dominance-submission-sadomasochist (BDSM) subculture has created a lexicon for understanding intimate relationships. It also serves as a microstudy in the power dynamics between two people. One overarching principle of the BDSM community is that consent requires communication.

In 1984 Paula Coughlin-Puopolo decided to join the navy and become a pilot. When she attended the U.S. Navy's annual Tailhook Conference in Las Vegas, she was subjected to the alcohol-induced "gauntlet." The gauntlet was a Navy ritual whereby male officers blocked elevators and trapped women in the hallway outside upper-floor hotel rooms. Coughlin-Puopolo recalled it vividly: "I felt that if I didn't make it off the floor, I was sure I was going to be gang raped.... I got attacked by a bunch of men that tried to pull my clothes off.... I fell down to the floor and tried to get out of the hallway, and they wouldn't let me out. They were trying to pull my underwear off from between my legs."[360]

Coughlin-Puopolo eventually took refuge in an empty hotel room, where she cried and tried to process what had happened. When she reported the incident to her superior the following morning, he retorted, "That's what you get for going down a hallway of a bunch of drunken aviators."[361]

When Coughlin-Puopolo filed a whistleblower complaint with the Navy, she was demoted and received "unrelenting pressure" to resign. She then went public, filing lawsuits and effectively ending her Navy career.[362]

Naturally, the whistleblower, not the perpetrator, was put on trial. According to defense attorneys, Coughlin-Puopolo had failed to call police or hotel security after the incident, implying she was lying. She was also portrayed as a "regular hotel slut." At one point, Coughlin-Puopolo met with Secretary of Defense Dick Cheney, who stated, "Because of your complaint, I have had to remove the secretary of the Navy."[363] Boys will be boys. In Cheney's mind, this was not the Navy's fault.

In the face of post-traumatic stress, court testimony, media scrutiny, unemployment, legal fees, and other financial and emotional distress, Coughlin-Puopolo took up yoga to avoid committing suicide.[364]

A Few Good Men

Humans are designed to propagate their species. The sex drive is therefore very powerful. A human who successfully abstains from his or her sex drive is even more powerful. World War II provided extreme insight on the nature of decision power in the context of sex. European women and children were left vulnerable as a result of martial law and virtually all men being called into military service. L.F. Filewood, a journalist and witness, described what he saw: "Young girls, unattached, wander about and freely offer themselves for food or bed.... Very simply, they have one thing left to sell, and they sell it."[365]

Russian soldier Aleksandr Solzhenitsyn also reported his experience: "For three weeks the war had been going on inside Germany, and all of us knew very well that if the girls were German, they could be raped and then shot."[366]

One German woman had a harrowing account of "twenty-three soldiers, one after the other. I had to be stitched up in a hospital. I never want to have anything to do with any man again."[367]

An anonymous U.S. soldier reported the following:

> "There are a few things about the Russian situation that I'd like to point out to you—the result of observations made during a two-and-a-half-months' stay in Austria.... The American soldier of all ranks has looted, and I speak only of things I have seen with my own eyes.... In the matter of rape, it is probable that more Russian soldiers have been guilty thereof than American, but much of the differential can be explained by the apparent predilection the Continental girl has for the American soldier with his K-ration, chocolate bars and cigarets [sic]..."[368]

According to U.S. Air Force Lieutenant Colonel Gerald F. Beane, "Rape presents no problem for the military police because a bit of food, a bar of chocolate, or a bar of soap seems to make rape unnecessary."[369] Colonel Beane's ignorance helps illustrate why hoarding life essential resources is an abuse of power.

In the case of consensual sex, each participant effectively has a vote. In the case of nonconsensual sex, decision power is centralized in the person controlling the options (resources). For example, one person's resources (physical force, a gun threat, other violence, money, a bar of chocolate, a bar of soap, and so on) are leveraged to overpower another person. When sex involves exchanges of life-essential resources, the decision to have sex is not consensual.

1985: USA Gymnastics

In 1985 Lawrence Nassar graduated from Michigan State University (MSU) with a degree in kinesiology. After completing medical school and becoming a professor at MSU, he decided to volunteer as a trainer and team doctor at several high schools and with USA Gymnastics, the agency governing women's gymnastics and the U.S. Olympic gymnastics team. In high school, Nassar learned he could touch girls if he spun himself as an unpaid volunteer providing free medical treatments to female coeds. As a doctor, Nassar gained even more power.[370]

In 1984 teenager Mary Lou Retton became the first U.S. Olympic gymnast to win an all-around gold medal. Kids everywhere suddenly had dreams of becoming Olympic gymnasts.[371] While millions of dollars in product endorsements started pouring into USA Gymnastics, so did the sexual assault complaints. Over the next thirty years, hundreds of gymnasts, some as young as six years old, made sexual assault allegations against Nassar.[372] Nassar repeatedly and unnecessarily inserted his ungloved fingers into vaginas during medical examinations. Nassar would spin his actions as valid "medical treatments" while spouting medical jargon.[373]

Deciding to Whistle Blow, Sarah Jantzi

At a competition in 2015, a coach overheard U.S. Olympic gymnastics hopefuls Maggie Nichols and Ali Reisman complaining about Nassar. Commendably, coach Sarah Jantzi blew the whistle, risking her career by reporting the accusations to parents and USA Gymnastics officials. Parents and athletes trusted Nassar and the USA Gymnastics coaches and officials. These coaches and officials had the power to decide whether an athlete made the U.S. Olympic Team.

In the case of Nichols, pursuing the sexual assault complaint may have cost her a spot at the 2016 Olympics.[374] USA Gymnastics had clear conflicts of interest. They were paying their attorneys top dollar to keep the allegations quiet.[375] They wanted to protect their brand and keep millions of dollars in commercial product endorsements rolling in ahead of the Olympics. USA Gymnastics officials decided that protecting gymnasts from a sexual predator was a lower priority than bringing in advertising revenue.

Indiana-based USA Gymnastics officials had a legal obligation to report child sex crimes. Instead, they approached the FBI with a bribe. USA Gymnastics officials successfully manufactured a conflict of interest for FBI decision-makers. USA Gymnastics CEO Stephen Penny asked FBI Special Agent in Charge (SAC) W. Jay Abbott, the highest-ranking FBI official in the state of Indiana, to meet him at a sports bar. Instead of offering evidence, Penny offered Agent Abbott a prestigious and well-paid position as a security consultant with the U.S. Olympic Committee. Without ever investigating the merits of the accusations against Nassar, Agent Abbott later suggested the FBI could issue a public statement favorable to USA Gymnastics.[376]

The parents of the athletes also had conflicts of interest. They did not want to miss an opportunity to attend the Olympics. For almost thirty years, parents of young athletes remained silent. They were effectively deciding that protecting their Olympic hopes was a higher priority than protecting future victims. Their silence was an abuse of power. The family of one gymnast who made the 2016 Olympic Team accepted a confidential $1.25 million settlement and signed a nondisclosure agreement with USA Gymnastics. This was how the problem perpetuated. When Nassar was finally arrested in 2016 by a team of female Michigan state law enforcement officials, investigators discovered more than 37,000 instances of child pornography on his computers.[377]

The Game within the Game

At about the same time as USA Gymnastics was figuring out how to dodge an FBI criminal investigation, the National Basketball Association (NBA) was too. NBA owners understood that it is more profitable to play superstars than to have them sit on the bench. Leagues like the NBA do not want superstars injured, given a night off, or benched for fouls. For a league like the NBA, where playoffs are a best-of-seven series, it is also more profitable for a series to go seven games instead of ending in a four-game sweep. Each game provides additional profit for the league, its owners, and the venue. As such, NBA officials are financially incentivized to fix the games to maximize profit. NBA official Tim Donaghy explained:

> "When you're hired in the league, you think everything is done based on the rules—how they are written in the rule book. You quickly learn, if you're gonna survive, that's not the way things are done. I started to understand the game within the game of the NBA, and I'll give you an example. I was in Philadelphia, and I'm refereeing Sixers-Bulls. [NBA executives] were cracking down on a spin move, and they wanted the officials to call traveling. And Michael Jordan does the spin move, and I make the call. Phil Jackson comes flying off the bench, and he starts giving me shit. And I say, 'Wait a second Phil, you know as well as I do that's

> the spin move that they are telling us to call.' And [Phil] said, '[NBA executives] may want that play called, but they certainly don't want it called on [Michael Jordan].' And [Phil] pointed at Jordan, who walked by and just stared at me. I got in the locker room, the other referee said, 'They want that call, but don't call it on [Jordan].' The way that [NBA Commissioner] David Stern structured the league, we as officials knew that it was better to treat the star players better than others. I just wanted to be the best referee and advance up the ladder. And I saw the way [officials] that were in the NBA finals handled the game. They didn't call fouls against the stars, and they were well respected. These people that pay thousands of dollars to sit courtside here, they didn't come here to see Kobe Bryant, LeBron James, Shaquille O'Neal sit on the bench. They came to see them play. The league's your boss, and you want them to think you're doing a good job because, if that's the case, you're on that playoff roster that comes out. With that is an enormous amount more money and respect."[378]

That's modern governance in a nutshell. The 99% get punished for "calling fouls" on the 1%. Each generation must learn how to play the game within the game.

After Donaghy agreed to wear a wire around NBA officials, executives, and owners, the FBI suddenly ended the investigation. According to FBI Special Agent Paul Harris, Stern offered him a higher-paying job with the NBA. Stern was never criminally investigated for bribery, and the NBA's FBI problem went away. The people in power avoided accountability while Donaghy went to prison.[379, 380]

CHAPTER 15

Quid Pro Quo

1986: Decisions Involving Hiring

Hiring decisions always involve a quid pro quo exchange of option power. The employer offers financial resources, such as a salary. The prospective employee offers human resources, such as labor.

After founding Miramax Films, Harvey Weinstein thrust himself on the Hollywood scene with several high-grossing independent movie productions. By the late 1980s Weinstein had the power to decide which actresses would be hired for Miramax movies. Some of the most attractive women in Hollywood would meet privately with Weinstein to vie for good-paying and high-profile acting roles in movies such as *Pulp Fiction*, *Scream*, and *Good Will Hunting*. Weinstein seized the opportunity to satisfy his personal interests.

Thousands of aspiring young actresses would avail themselves to casting calls. A hiring decision need not be made. In fact, no movie was technically needed. The propaganda was easy. Weinstein could lie, selling the opportunity as a lead role in an upcoming blockbuster film and possibly bolstering it with, "You'll be working alongside an Academy Award winner." The idea of working with elite actors was propaganda intended to influence a decision. Weinstein could make future promises he had no intention of fulfilling.

Actresses would sometimes meet with Weinstein alone in hotel rooms. Well-paid Miramax employees, having their own conflicts of interest, would dutifully escort the unsuspecting actresses in and out. Each actress would leverage her own variety of resources (experience, charm, persuasiveness, a reduced salary, a short skirt, cleavage, or other forms of sex appeal) in hopes of securing a favorable decision. Some actresses benefited from these power exchanges. Others did not.

According to actress Asia Argento, Weinstein invited her to a Miramax party at a luxurious hotel. Argento was then invited to an after-party in Weinstein's hotel room. That was red flag number one. A Miramax employee led Argento to the room, but the room was empty. That was red flag number two. The employee told Argento they had arrived early, and then left the room—red flag number three. Regardless, Argento was determined to pursue her acting dreams. Weinstein was increasing his power over her through a social mechanism, spun as an after-party, which minimized accountability by eliminating witnesses. Simply put, Weinstein controlled the setting.

Weinstein walked in wearing a robe, holding a bottle of lotion, and asking Argento for a massage. As Argento described it, Weinstein "terrified me… he was so big." Weinstein was six feet tall and weighed more than 200 pounds. He had the physical resources to overpower his much smaller victim. Weinstein pulled up Argento's skirt, forced her legs apart, and licked Argento's vagina as she repeatedly told him to stop. Argento recalled feigning enjoyment, believing it was the only way the assault would end.[381]

When accusations of rape were made against Weinstein, he had the power to make them disappear. Weinstein could end an accuser's career with one phone call to any other Hollywood executive, or he could purchase her communication power with hush money. He could legally overpower an accuser by hiring attorneys to ruin the reputation and credibility of an actress.[382] Weinstein went so far as to hire the Israeli intelligence corporation BlackCube (former Mossad agents) to silence his accusers.[383] Weinstein was so wealthy that he could afford the substantial cleanup costs associated with his bad behavior.

Actress Rose McGowan was another Weinstein victim. When McGowan told the head of Amazon Studios that Weinstein had raped her, she was ignored.[384] One whistleblower alone was never enough. Stopping Weinstein required the collective communication power of several victims. In an age of centralized corporate mass media, leveraging journalists was also critical. After interviewing dozens of victims, New York Times investigative journalists Jodi Kantor and Megan Twohey blew the whistle on Weinstein. Thirty years after the first accusations of rape, Weinstein was finally arrested.

1987: Decisions Involving the Environment

By the 1980s, oil dependency was putting the financial security of the 1% ahead of the environmental security of the 100%. As a result, the National Aeronautics and Space Administration (NASA) was summoned to testify before the U.S. legislative branch regarding climate change. NASA sent its top climate scientist, James Hansen, an expert in atmospheric physics. In summary, Hansen testified to the following:

> (1) By burning fossil fuels, particularly oil, CO_2 is being released into the atmosphere.
> (2) Increased atmospheric CO_2 is causing air pollution and smog.
> (3) Increased atmospheric CO_2 is causing increased temperatures worldwide (the greenhouse effect).
> (4) Increased temperatures are melting ice caps and increasing sea levels.
> (5) Increased temperatures are causing heat waves and adding water vapor to the atmosphere.
> (6) Adding water vapor to the atmosphere will increase the frequency and intensity of future hurricanes and floods.

As Hansen put it, "It is already happening now."[385]

Self-regulation had allowed oil corporations to conceal evidence of environmental harm caused by their products.[386] In response to public outcry over global warming, Republicans began covertly censoring information involving climate change. Press releases from federal agencies such as NASA and the

Environmental Protection Agency (EPA) were edited to make global warming seem like less of a threat.[387] In an obvious conflict of interest, they were protecting their oil industry allies and campaign donors.[388]

1988: The CIA Decides the U.S. Presidential Election

In 1988 CIA Director George H.W. Bush became the forty-first U.S. president. His rise to power came after a string of miraculous election victories in numerous states. It was the first time that electronic voting machines were used in a U.S. presidential election. Votes were counted using centralized computers.

During the 1988 New Hampshire Republican primary, candidate Bush trailed by eight percentage points in public polls. However, when the Republican staff of New Hampshire Governor John Sununu, a computer engineer, counted the electronic votes, Bush beat his opponent by nine percentage points. According to analysts, the seventeen-point swing was a statistical impossibility.[389] After becoming president six months later, Bush provided Governor Sununu with a quid pro quo, appointing him as White House chief of staff.

During the former CIA director's first term as president, Soviet Russia collapsed. Ironically, the CIA did not see it coming. CIA Director Stansfield Turner responded by saying, "We should not gloss over the enormity of this failure."[390]

Similarly, U.S. Senator Pat Moynihan concluded, "For a quarter of a century, the CIA has been repeatedly wrong about every major political and economic question entrusted to its analysis."[391]

CIA officials had successfully proven that humans can't see the future.

CHAPTER 16

Criminalization

1989: Decisions Involving a War on Drugs

By 1989 Republican Party officials were pivoting to a new war. Despite the sudden end of the Cold War, CIA funding remained at an all-time high. As a CIA agent, President Bush had helped globalize heroin, cocaine, LSD, and other drugs. As a U.S. House representative (1967–1971), U.S. vice president (1981–1989), and U.S. president (1989–1993), the former CIA director helped to propagate the idea of a "war on drugs."[392]

As a House representative, Bush helped President Nixon push the Controlled Substances Act (CSA) through the legislative branch. The CSA criminalized marijuana, heroin, cocaine, LSD, and other drugs. By making these drugs illegal, the U.S. supply was decreased, driving up the price.[393, 394] Because the CIA controlled and protected several international drug cartels, it was in a position to heavily profit off the books.[395, 396]

During a public speaking event as president, Bush told the following dramatic story:

> "Not long ago, I read a newspaper story about a little boy named Dooney, who, until recently, lived in a crack house in a suburb of Washington, DC. In Dooney's neighborhood, children don't flinch at the sound of gunfire. And when they play, they pretend to sell to each other small white rocks that they call crack. Life at home was so cruel that Dooney begged his teachers to let him sleep on the floor at school. And when asked about his future, six-year-old Dooney answers, 'I don't want to sell drugs, but I'll probably have to.'"[397]

The beauty of propaganda is that it is easy to manufacture. As the Nazis figured out, all it requires is a little imagination and a lot of communication power (centralized media). Like any six-year-old, Dooney had apparently put a lot of thought into his future employment options and decided that he will "probably" have to sell drugs to make ends meet.

1990: Decisions Involving Law Enforcement

During President Nixon's first year in office, Bush served as ambassador to the United Nations and worked closely with Nixon's Domestic Policy Chief John Erlichman. While Bush helped the CIA install wiretaps at the United Nations, Erlichman helped

Nixon lay the groundwork for a race war spun as a drug war.[398] Erlichman described it in this way:

> "The Nixon campaign in 1968, and the Nixon White House after that, had two enemies: the antiwar left and black people. You understand what I'm saying? We knew we couldn't make it illegal to be either against the war or black, but by getting the public to associate the hippies with marijuana and blacks with heroin, and then criminalizing both [drugs] heavily, we could disrupt those communities. We could arrest their leaders, raid their homes, break up their meetings, and vilify them night after night on the evening news. Did we know we were lying about the drugs? Of course we did."[399]

Despite the end of the Cold War, CIA officials had created a new way to continue routing U.S. tax dollars to the military-industrial complex. As president, Bush had directed Congress to fund the war on drugs. The war on drugs enabled U.S. weapon corporations to militarize U.S. federal law enforcement agencies as well as local police.[400]

The war on drugs also gave way to the prison-industrial complex, which drove up the U.S. incarceration rate and enabled expansion for private prison corporations. The process was highly profitable and allowed predominantly white politicians and police forces to incarcerate a disproportionate number of nonwhites.[401]

1991: Decisions Involving Employment

As a way to covertly fund its bad behavior, the CIA secretly empowered international drug cartels to smuggle cocaine into the states.[402] One such smuggler was introduced to an entrepreneurial twenty-year-old named Ricky Ross, a high school dropout from South Central Los Angeles. Ross capitalized on the opportunity by using members of two African American street gangs, the Bloods and the Crips, to sell the cooked cocaine as "crack." Crack rocks were more marketable than cocaine powder because they transported easily, looked like candy, and offered a more intense high.[403, 404]

Recruiting exploded for the Bloods and Crips. Young black men could not find better pay. They further secured their trade by investing in semiautomatic weapons. At one point, Ross had more than 1,000 employees selling approximately 500,000 crack rocks each day. He quickly expanded to Texas, New York, Ohio, Pennsylvania, Maryland, Missouri, Indiana, and Washington.[405] Each week, Ross purchased roughly 400 kilograms of Nicaraguan cocaine. Between 1980 and 1988, Ross grossed approximately $1 billion annually. Communities in the Los Angeles area and across the U.S. became ravaged by a crack epidemic.[406]

CHAPTER 17

Violence

1992: Decisions Involving Race

In 1992 an intoxicated Rodney King was pulled over for speeding by white Los Angeles police officers. King, an African American, stopped his car in front of an apartment complex whereby residents witnessed the arrest. Technology allowed the whole world to watch. An apartment resident with a camcorder videotaped four police officers swarming a docile King and beating him with batons. Eight additional business-as-usual police officers stood by as King suffered a fractured skull that resulted in brain damage.[407]

The video, once broadcast, communicated indisputable accountability information. It communicated what every African American already knew: They are treated unequally and as if their lives don't matter. After the trial, when four of the officers were found not guilty of assaulting King, parts of Los Angeles erupted.

Like previous race riots, dozens of people were killed or arrested.

Race discrimination was not unique to African Americans. One decade earlier, Vincent Chin, a Chinese American, was celebrating his bachelor party in Detroit, Michigan. At some point, a fight broke out between Chin and two unemployed white male autoworkers. According to witnesses, one of the men told Chin, "It's because of you little motherfuckers that we're out of work."[408]

One man was holding Chin while the other beat him to death with a baseball bat. The two men were charged with manslaughter, receiving three years' probation and no jail time. According to Judge Charles Kaufman, a Japanese prisoner of war during World War II, "These weren't the kind of men you send to jail."[409]

Policing and adjudicating are subprocesses of accountability power. When police and judicial powers are primarily centralized in white Christian males, lack of accountability over that group is inevitable. Nonwhites disproportionately become the targets of criminal investigations. Nonmales suffer from unequal pay and unequal credibility in retaliation matters. Non-Christians disproportionately become the target of terrorism investigations.

Democracy requires equal accountability power for all group members. When group members are on the wrong end of power inequality, they become angry and violent. By U.S. Constitutional standards, they progress from the First Amendment to the Second

Amendment, from free speech to bearing arms. It is human nature.

1993: Decisions Involving Bearing Arms

In 1993 Al-Qaeda put its name on the map when it bombed U.S. soldiers in Yemen, killing two people. Five years later, more than 200 people were killed and over 4,000 injured from Al-Qaeda bombings at two U.S. embassies in east Africa. During the criminal trial of the bombings, U.S. government prosecutors exposed far more accountability information than they hoped for. It became apparent that people in the Middle East were angry. Many hated the U.S.—for rigging their elections, bribing their decision-makers, and assassinating their most beloved leaders—and they were now bearing arms and fighting for their freedom.

Essam Al-Ridi, a U.S. citizen and pilot, publicly revealed that, under Operation Cyclone, the CIA had secretly spent up to 600 million U.S. tax dollars each year supplying small arms, antiaircraft weapons, and surface-to-air missiles to the mujahedin. The mujahedin were mercenaries covertly hired to fight the Russians. One of the mujahedin was a Saudi Arabian named Osama Bin Laden. At the expense of U.S. taxpayers, Bin Laden and other mujahedin spent the 1980s becoming experts in terror. When the mujahedin worked for the U.S., they were labeled freedom fighters. When the mujahedin worked against the U.S., they were labeled terrorists.

Through the U.S. embassy bombings trial, due process provided Americans with priceless accountability information. Al-Ridi's testimony provided evidence that the CIA was cultivating more terror than it was preventing. Yet that accountability information drew a collective yawn from the American public. Perhaps Americans were distracted by cultural noise, such as the Backstreet Boys, *Sex and the City*, Harry Potter, and PlayStation. Accountability was not a priority. Al-Qaeda was about to change that.[410, 411, 412]

In 1993 Al-Qaeda also attacked New York's World Trade Center (WTC), killing six people. Eight years later, on September 11, 2001 (9/11), both towers of the WTC were destroyed by Al-Qaeda operatives armed with American jetliners. More than 2,000 people were killed. Osama Bin Laden subsequently communicated the following accountability information to the American people:

> "The events that affected my soul in a direct way started in 1982 when America permitted the Israelis to invade Lebanon and the American Sixth Fleet helped them in that. This bombardment began and many were killed and injured, and others were terrorized and displaced. I couldn't forget those moving scenes, blood and severed limbs, women and children sprawled everywhere. Houses destroyed along with their occupants, and high rises demolished over their residents, rockets raining down on our home without mercy.... And as I looked at those demolished towers in Lebanon, it entered

> my mind that we should punish the oppressor in kind and that we should destroy towers in America in order that they taste some of what we tasted and so that they be deterred from killing our women and children. And that day, it was confirmed to me that oppression and the intentional killing of innocent women and children is a deliberate American policy.... This is the message which I sought to communicate to you in word and deed, repeatedly, for years before September 11. And you can read this, if you wish, in my interview with Scott [MacLeod] in *Time* magazine in 1996, or with Peter Arnett on CNN in 1997, or my meeting with John Weiner in 1998."[413]

Subsequently, the U.S. invasion of Afghanistan was never about fighting "Islamic terrorism." That was propaganda. Islamic terrorism was a pretext—a threat manufactured to ensure oil resources could be secured through a perpetual war on terrorism. U.S. Senator Chuck Hagel later confessed, "People say we're not fighting for oil. Of course we are. They talk about America's national interest. What the hell do you think they're talking about? We're not there for figs."[414]

Federal Reserve Chairman Alan Greenspan also confessed, "I am saddened that it is politically inconvenient to acknowledge what everyone knows: the Iraq war is largely about oil."[415]

U.S. Army General John Abizaid unequivocally admitted, "Of course it's about oil. it's very much about oil, and we can't really deny that."[416] Abizaid added,

"We must adopt, as a matter of nation security policy, a way to reduce our dependency on Middle Eastern oil."[417]

The concept of national security is great when it actually secures a nation. When it secures the financial interests of oil companies and the 1%, it's propaganda.[418]

Following 9/11, Osama Bin Laden pronounced the following:

> "All that we have to do is to send two mujahedin to the farthest point East, to raise a piece of cloth on which is written 'al-Qaida,' in order to make the generals race there; to cause America to suffer human economic and political losses without their achieving for it anything of note other than some benefits to their private companies. This is in addition to our having experience in using guerrilla warfare, and the war of attrition to fight tyrannical superpowers as we did, alongside the mujahedin, bled Russia for ten years, until it went bankrupt and was forced to withdraw in defeat. All Praise is due to Allah. So we are continuing this policy in bleeding America to the point of bankruptcy." [419]

The wars in Iraq and Afghanistan cost U.S. taxpayers an estimated $5 trillion, raising the U.S. national debt to $20 trillion (2017). That money went right into the pockets of the military-industrial complex and the U.S. 1%.[420]

Maintaining Centralized Power over Global Energy Resources

Before becoming the U.S. vice president in 2001, then-Halliburton CEO Dick Cheney made the following disclosure in a private speech to the Institute of Petroleum:

> "By 2010, we will need on the order of an additional fifty million barrels a day. So where is the oil going to come from? Governments and national oil companies are obviously controlling about 90% of the assets. Oil remains fundamentally a government business. While many regions of the world offer great oil opportunities, the Middle East, with two-thirds of the world's oil and lowest cost, is still where the prize ultimately lies. Even though companies are anxious for greater access there, progress continues to be slow."[421]

Instead of engineering alternative energy options, the U.S. 1% decided to engineer a war. According to former Lockheed Martin Director Ben Rich, more democratic energy options are already in the possession of the U.S. 1%. Such technologies are hidden in so-called black projects.[422, 423, 424]

Operation Majestic 12

Imagine a person taking a cross-country road trip. There would be a need to stop for resources such as fuel, water, or a place to rest. The same principle would hypothetically apply to extraterrestrial travel. How would an extraterrestrial pay for local resources? Extraterrestrial currency would likely have no value on Earth; however, information related to sustainable energy, antigravity, propulsion, stealth, supercomputers, and weapons of mass destruction would be highly valuable on a planet lacking such technologies.[425]

In 1984 an anonymous whistleblower left a package at the front door of Los Angeles television producer Jaime Shandera. Inside that package was microfilm. Once developed, the microfilm produced images of documents referencing a U.S. government project called Majestic 12. According to the images, Majestic 12 was a secret group of twelve wealthy white Christian males assembled in 1947 to investigate extraterrestrial activity, such as the saucer that crashed in Roswell, New Mexico, that year. Majestic 12 was vested with centralized accountability power over all information resources involving extraterrestrials.[426]

Consisting of unelected U.S. government officials, the original members of Majestic 12 were the following people:

MJ-1: The third CIA director, Navy Vice Admiral Roscoe Hillenkoetter
MJ-2: National Defense Research Committee Chairman Vannevar Bush (Raytheon founder)
MJ-3: The first secretary of defense, Navy Lieutenant James Forrestal
MJ-4: Air Force General Nathan Twining

MJ-5: The second CIA director, Air Force General Hoyt Vandenberg
MJ-6: Army electrical engineer and biophysiologist Detlev Bronk
MJ-7: Navy Commander Jerome Hunsaker
MJ-8: The first CIA director, Navy Admiral Sidney Souers
MJ-9: Army attorney Gordon Gray
MJ-10: Navy astronomer and astrophysicist Donald Menzel
MJ-11: Army Commander Robert Montague (Sandia Base, New Mexico)
MJ-12: Navy geophysicist Lloyd Berkner[427]

NASA astronaut Edgar Mitchell described it this way:

> "Yes, there have been [extraterrestrial] visitations. There have been crashed craft. There have been material and bodies recovered. And there is some group of people somewhere that may or may not be associated with government at this point, but certainly were at one time, that have this knowledge. They have been attempting to conceal this knowledge or not permit it to be widely disseminated."[428]

The Roswell crash in 1947 occurred roughly 100 miles from Earth's first modern nuclear weapons test site. It would appear planet-destroying technology attracted the attention of extraterrestrials.[429]

One reason for secrecy surrounding extraterrestrials was technology. The technology in one extraterrestrial aircraft, like at Roswell, would have been worth trillions of dollars. It would enable instant economic growth for whatever country controlled that technology. Technological resources offered centralized option power. During an alumni speech at the University of California Los Angeles, former Lockheed Martin Skunk Works division CEO Ben Rich put it this way: "We already have the means to travel among the stars, but these technologies are locked up in black projects, and it would take an act of God to ever get them out to benefit humanity. Anything you can imagine, we already know how to do it."[430]

Humans are now communicating at a rate never before seen in Earth's history. This has afforded government employees the opportunity to not only whistle blow but to whistle blow with evidence in hand. Just as Shandera's microfilm held dozens of Majestic 12 documents, transportable digital memory can now hold millions of documents. Using social mechanisms like the Internet, WikiLeaks, or the International Consortium of Investigative Journalists (ICIJ), evidence can be wirelessly decentralized—transmitted to the entire world—in a matter of seconds. Technology enables accountability.

For centuries, covert government entities all over the world have operated under the assumption that they perpetually could exist in secret. These entities believed they could overthrow governments, bribe politicians, assassinate world leaders, manipulate economies, and control resources without anyone finding out. After all, secrecy was easy: microfilm, photocopiers, transportable digital memory, computers, the Internet, and so on, did not yet exist. Until the end of the twentieth century, government decisions were rarely documented in writing. There was no evidence to point to in the name of accountability. Other than the perpetrators, humankind never knew the truth.

After World War II, the CIA became one of those covert government entities. CIA officials never expected to be caught, let alone held accountable. Whistleblowers have been the biggest factor in this accountability process. In a 1% system, whistleblowers are proving indispensable to the process of decentralizing power and balancing power inequalities.

Leveraging Advanced Technology: Ellsberg and the Pentagon Papers

During the Vietnam War, technology helped to do something the U.S. legislative and judicial branches have consistently failed to do: hold the U.S. executive branch accountable. Department of Defense analyst and whistleblower Daniel Ellsberg photocopied top secret documents and delivered them to the New York Times. The Pentagon Papers contained more than 7,000 pages of a Pentagon analysis into how four U.S. presidential administrations successfully lied to the American people by fabricating evidence needed to justify U.S. military involvement in Vietnam.

Following the assassination of JFK, LBJ assumed the role of president. Within a year, LBJ cited the attack of U.S. Navy ships by the North Vietnamese as justification for a formal U.S. invasion of Vietnam. LBJ claimed the attack occurred in the Gulf of Tonkin. U.S. Navy whistleblower James Stockdale was the commander assigned to the Gulf of Tonkin at the time of the alleged attack. According to Commander Stockdale, LBJ was lying: "I had the best seat in the house to watch that event… our destroyers were just shooting at phantom targets."[431]

The American people were clueless. The U.S. 1% tricked the U.S. 99% into believing they were under attack. There were no Vietnamese ships in the Gulf of Tonkin at the time. Despite violent public protests, the Vietnam War lasted for almost two decades. During that time, the U.S. legislative and judicial branches never lifted a finger. The U.S. political party system controlled all three branches. Government accountability, pursuant to U.S. Constitutional social mechanisms like checks and balances, was short-circuited.

Years later, U.S. Secretary of Defense Robert McNamara confirmed LBJ had lied to the American people. The Gulf of Tonkin incident was a false flag. By leveraging technology and asserting his rights to communication and accountability, Ellsberg may have ended the Vietnam War. Ellsberg's actions were a technodemocratic revelation utilizing photocopier technology to communicate more than 7,000 pages of accountability information to the American people. This momentary illumination by a lone government whistleblower was not appreciated by the U.S. 1%.[432, 433]

CHAPTER 18

Coconspirators

The White House Plumbers and the Run-up to Watergate

In response to Ellsberg's "leak," then-President Nixon secretly created the White House Plumbers. The Plumbers were a group of Republican henchmen consisting of former FBI and CIA agents. They were tasked with covertly stopping information leaks and discouraging whistleblowing.

The Plumbers burglarized the offices of Ellsberg's psychiatrist, looking for information to publicly discredit Ellsberg's mental health. They also attempted to poison Ellsberg with LSD before a public speaking event. Both operations failed, as did the criminal prosecution of Ellsberg under the U.S. Espionage Act.

At Ellsberg's trial, Judge William Byrne declared, "The totality of the circumstances of this case, which I have only briefly sketched, offends a sense of justice.... The bizarre events have incurably infected the prosecution of this case."[434]

The courtroom erupted in cheers as Ellsberg went free. Several years later, Judge Byrne made a startling confession involving a conspiracy to imprison Ellsberg. During Ellsberg's trial, President Nixon had secretly offered Judge Byrne the position of FBI director as quid pro quo for a conviction of Ellsberg.[435, 436]

The Saturday Night Massacre

During his first term as president, Nixon's Republican Committee to Re-Elect the President (CREEP) paid the Plumbers to break in and wiretap the offices of the Democratic National Committee (DNC). Local police arrested the Plumbers inside the DNC's Watergate office. Because CREEP had laundered campaign donations through a Mexican bank account to fund the break in, Nixon immediately ordered his CREEP coconspirators to obstruct the FBI's Watergate investigation and cover up the crime.[437]

Nixon's direct involvement in the Watergate break-in may never have been brought to light but for an anonymous FBI whistleblower codenamed Deep Throat by Washington Post reporters. Recording technology would then prove critical in holding Nixon accountable for the Watergate burglary.[438] Thanks to Deep Throat, a special prosecutor was appointed by the legislative branch to investigate Nixon. The special prosecutor subsequently subpoenaed Nixon, demanding copies of incriminating White House tape recordings. Nixon's resulting waterfall of employment retaliation was referred to by the press as the Saturday Night Massacre.[439, 440]

On the Saturday night after the tape recordings were demanded by the special prosecutor, President Nixon ordered his attorney general to fire the special prosecutor. The attorney general refused, and then resigned. The president then ordered his deputy attorney general to fire the special prosecutor. The deputy attorney general refused, and then resigned. President Nixon then ordered Solicitor General Robert Bork to do the deed. Bork complied, later admitting he too had considered resigning so as to avoid being perceived as "a man who did the President's bidding to save my job."[441]

So why did Bork do the president's bidding? In exchange for Bork's decision to fire the special prosecutor, President Nixon had secretly promised to abuse his decision power and appoint Bork to the next open seat on the U.S. Supreme Court, quid pro quo. When a seat opened on the Supreme Court in 1976, Bork was appointed by Republicans.[442]

Unprosecuted Republicans: Nixon, Rove, Scalia, Rumsfeld, Cheney, and Bush

By 1974 the Watergate investigation had resulted in criminal convictions and plea deals for numerous Republican Party coconspirators, including two attorneys general, an assistant attorney general, the White House chief of staff, two White House counsels, the special counsel to the president, the Securities and Exchange Commission (SEC) chairman, and one of the Watergate burglars, an active CIA agent named Eugenio Martinez. Martinez was eventually pardoned by Republicans.[443]

While a Republican Party network of more than seventy officials was charged with Watergate crimes, several notable members of the Nixon administration avoided any charges, including CREEP associate Karl Rove, General Counsel Antonin Scalia, Counsel to the President Donald Rumsfeld, White House Staff Assistant Dick Cheney, and U.S. Ambassador to the United Nations George H.W. Bush. Only one month after resigning, Nixon was pardoned by his former vice president and Republican colleague, President Gerald Ford. The pardon allowed Nixon to avoid criminal prosecution.[444]

After Watergate, Secretary of State Henry Kissinger warned Ford, "When the FBI has a hunting license into the CIA, this could end up worse for the country than Watergate."[445]

In hopes of avoiding criminal prosecution, CIA officials had every incentive to perpetually install CIA agents in the offices of both the U.S. president and the FBI.

The Watergate Whistleblower: FBI Agent Mark Felt

Over thirty years after the Watergate scandal broke, Deep Throat was able to reveal his identity with little retaliation. Retired FBI Assistant Director Mark Felt, the FBI's second in command during the Watergate investigation, admitted he was Deep Throat. According to Felt, then FBI Director L. Patrick Gray, who had been appointed by Republicans to replace Hoover, was secretly passing FBI investigative findings directly to the perpetrator, Nixon. In exchange for making him the next FBI director, Gray had agreed he would leak FBI information resources to Nixon, quid pro quo.[446]

At the same time, Republican Party contacts within the CIA conspired to send false leads to the FBI's Watergate investigators. Although he feared for his career and his life, Felt covertly blew the whistle and communicated accountability information to the American people, via the Washington Post.[447]

Before being forced to resign, Nixon appointed four justices to the U.S. Supreme Court: Warren Burger, Harry Blackmun, Lewis Powell, and William Rehnquist. Because Nixon appointed them, it is not unreasonable to suspect some form of bribery or quid pro quo was involved. All four Nixon-appointed justices decided *Buckley v. Valeo* in 1976. All four ruled that campaign donation limits are unconstitutional, opening the door to an era of unprecedented campaign donations and overt bribery.[448]

Due to Nixon-era Republicans having a history of appointing U.S. Supreme Court Justices based on quid pro quo secret agreements, greater accountability efforts may be needed to identify ongoing conflicts of interest involving current Justices Samuel Alito, John Roberts, and Clarence Thomas. All three were appointed by a former member of the Nixon administration.

Undermining Accountability: Political Party Loyalty

Immediately following the Watergate scandal, the legislative branch passed the Privacy Act, which amended the Freedom of Information Act (FOIA). Ordinary Americans could obtain copies of previously classified documents. Even with the Watergate scandal fresh on the minds of the American people, the legislative branch needed an overriding two-thirds vote to pass the Privacy Act after then-President Ford vetoed it. Ford was apparently persuaded to veto the Privacy Act by Chief of Staff Donald Rumsfeld, Deputy Chief of Staff Dick Cheney, and Assistant Attorney General Antonin Scalia.[449, 450]

In 1975, Idaho Senator Frank Church put the executive branch in check. The Church Committee was formed to congressionally investigate secret government operations. In closed-door hearings, CIA officials admitted they engaged in election rigging but insisted they had not rigged U.S. elections.[451]

Senator Church concluded that the CIA had become a "rogue elephant" running up a tab on U.S. taxpayers without any accountability. Led by Rumsfeld, Republicans blocked the Church Committee from dismantling the CIA. The Church Committee was then dissolved, keeping power centralized in the CIA.[452]

Although an FBI whistleblower had provided the check, Republicans prevented the balance. Through political party loyalty, the CIA had dodged the checks-and-balances bullet of the U.S. Constitution.

1994: Agent Vladimir Putin Becomes the KGB's Savior

Following the collapse of the Soviet Union, the CIA successfully installed Boris Yeltsin as Russia's first president.[453]

After the KGB was publicly dissolved and replaced by the FSB, a siloviki—a network of former KGB agents and assets coconspiring with Russian politicians—remained.[454] The siloviki needed a way to covertly replenish their financial resources and retake control of the Russian government.

At the time, former KGB agent Vladimir Putin was working for the mayor of Saint Petersburg, Russia.[455] Putin began exporting over $100 million in city-owned precious metals in exchange for humanitarian aid. As described by city legislator Marina Salye, "There was no food in the city at all.... There was no money. Barter was the only way. Say, metals for potatoes and meat."[456]

As it turned out, the beneficiaries of the precious metals were siloviki companies. Humanitarian aid never arrived. The siloviki converted the precious metals into cash for future covert operations. Using cash kickbacks, Putin purchased a villa in Spain.[457] When local Saint Petersburg politicians attempted to criminally investigate Putin, the siloviki shielded him from prosecution.[458]

1995: The Russian Oligarchs

By 1995, President Yeltsin was privatizing Russian state-owned resources to, among other things, fund his reelection in 1996. Russian resources such as land, natural gas, and oil were effectively sold off. Behind the scenes, both the siloviki and the CIA separately conspired to rig the process. The siloviki successfully centralized the resources in a small group of billionaires called the Russian oligarchs.

Harvard Director of Russian Studies Marshall Goldman described how it was done: "Ownership of some of Russia's most valuable resources was auctioned off by oligarch-owned banks under a scheme called 'Loans for Shares.' Although they were supposedly acting on behalf of the state, the bank auctioneers rigged the process—and in almost every case ended up as the successful bidders."[459]

Most of the assets of the former Soviet Union were dissolved. Undocumented handshake agreements were common.[460] Boris Berezovsky, for example, acquired Sibneft Oil Company for only $1 million. Sibneft was worth roughly $3 billion.[461] Similarly, Mikhail Khodorkovsky acquired 78% of the shares in Yukos Oil Company for only $310 million. Yukos was worth roughly $5 billion.

1996: Deciding the Russian Presidency, the CIA

In 1996, President Yeltsin was running for reelection. He was fifth in early 1996 presidential polls but was about to be rescued by the CIA.[462] American campaign advisors for California Governor Pete Wilson and President Bill Clinton were secretly smuggled into the Kremlin's President Hotel in downtown Moscow. Under Russian campaign finance laws, candidates could only spend $3 million. According to various accounts, Yeltsin conspired to take secret campaign donations estimated at $700 million. The money was laundered and routed into the American equivalent of super PACs.[463]

For months, Yeltsin's American campaign advisors covertly orchestrated an American-style presidential campaign through propaganda targeting young Russian voters. Well-paid Russian celebrities were suddenly endorsing Yeltsin on MTV-style "choose or lose" shows. Russian rock stars were paid to go on an "our president" tour. Acne cream was marketed as "for Yeltsin." Hotel reservations were mysteriously cancelled for Yeltsin's opponents. To secure fake endorsements, bribes were secretly paid to established politicians. Yeltsin eventually won the 1996 Russian presidential election by a landslide.[464]

1997: Deciding the Russian Presidency, the KGB

By 1997 the siloviki had seen enough.[465] They wanted to install a KGB asset as Russian president—permanently. One candidate was Putin. To make Putin appear more qualified for the office of the president, the siloviki went to work on propaganda. To improve Putin's public image, they conspired for him to receive a fraudulent Ph.D. in economics. Putin's thesis was plagiarized using a 1978 textbook written by a professor at the University of Pittsburgh.[466]

As a false flag, the FSB covertly bombed apartment buildings in three Russian cities, killing more than 300 people. The FSB, which was also empowered with investigating the attacks, blamed Chechen terrorists.[467] As part of Putin's rise to the presidency, siloviki media assets propagated the idea that the former KGB agent was a unifier and patriot who could defend Russia from Chechen terrorists once he became president.[468]

When FSB agent and whistleblower Alexander Litvinenko publicly revealed the FSB was in fact responsible for the Russian apartment bombings, the siloviki had him poisoned to death using Polonium-210.[469]

1998: Deciding to Politically Assassinate Russian President Boris Yeltsin

While Putin was rapidly ascending from political advisor to FSB director to prime minister, FSB agents were collecting kompromat on CIA assets. One such CIA asset was Prosecutor General Yury Skuratov. Skuratov was the Russian equivalent of the U.S. attorney general and a 2000 presidential candidate. FSB agents entrapped and videotaped Skuratov in bed with child prostitutes. Putin then hand delivered the videotape to a Russian public television station.[470] After being politically compromised, Skuratov was no longer a viable presidential candidate.

With Skuratov out of the way, President Yeltsin was next. FSB agents conspired to create criminal charges against Yeltsin's family members. They had accepted bribes from government contractors and were now facing prison time. In exchange for not prosecuting his family members, Yeltsin agreed to resign as president. Under Russian constitutional law, Prime Minister Putin automatically became president. Putin never received a single vote.[471]

Russian oligarchs who previously resisted the siloviki now feared for their lives. As president, Putin used oligarch Mikhail Khodorkovsky to set an example. Khodorkovsky was imprisoned for fraud, embezzlement, money laundering, and tax evasion.[472] Oligarchs had option power (in the form of financial resources) while Putin had accountability power (through control of the FSB and law enforcement agencies). The oligarchs subsequently wanted to avoid any form of assassination. Quid pro quo, they became a new source of covert funding for Putin and the siloviki.[473]

1999: Decisions Involving Government Accountability

Political parties have no place in a democracy. They create conflicts of interest. Simply put, party politicians have personal interests in party loyalty that conflict with public interests in accountability.

For example, in Russia, Putin's political party has since enacted legislation to ensure Putin remains president until 2036.[474] Similarly, the Watergate scandal showed us how far U.S. political parties would go to protect their own. During Watergate, more than seventy Republicans were charged with crimes, including burglary, fraud, money laundering, warrantless wiretaps, bribery, and racketeering.[475] Both countries demonstrate why political parties are a threat to both national and global security.

Following the Watergate scandal, the U.S. legislative branch passed the Ethics in Government Act (EGA) in 1978.[476] The EGA was drafted such that it would automatically expire after twenty years. It enabled the legislative branch to perpetually and proactively investigate the executive branch. Naturally, this power was abused by political parties. Between 1978 and 1998, opposing political parties used the EGA to impeach every single U.S. president.[477]

By 1999, President Bill Clinton, the subject of seven EGA investigations, was acquitted by the legislative branch during an impeachment trial over the Monica Lewinsky scandal. Recognizing that political parties were abusing the EGA by attempting to assassinate each other politically, the legislative branch decided to let the EGA expire. Rather than abandon the political party system, the U.S. legislative branch decided to abandon one of its few remaining government accountability mechanisms.[478]

PART IV

ACCOUNTABILITY POWER

CHAPTER 19

Freedom Fighters

2000: Holding the U.S. Accountable

If there is no accountability, there is no democracy. With a half-century of hindsight, it is apparent the CIA covertly serves the interests of a centralized few. The goals of preventing another Hitler and maintaining secrecy over weapons technology were understandable. However, the corresponding centralization of power was clearly abused by the U.S. 1%. Following World War II, the U.S. Gross National Product (GNP) more than doubled every twenty years, drastically surpassing every other nation on Earth.[479] On one hand, by secretly centralizing global power, the CIA may have prevented World War III. On the other hand, the CIA may have already started it. The CIA has provoked a long list of enemies.

In 2000 George W. Bush (GWB) was inaugurated as the forty-third U.S. president. GWB was the son of the forty-first U.S. president, former CIA Director George H.W. Bush. For decades, U.S. infrastructure corporations, via the CIA, obtained multimillion-dollar infrastructure contracts from corrupt foreign decision-makers. In the case of the Middle East, U.S. engineers, instead of Middle Eastern engineers, were brought in to fulfill contracts.[480]

In response, several unemployed Middle Eastern engineers decided to readapt their skills by infiltrating the U.S. and counterattacking. Because nonviolent communications by Osama Bin Laden had failed, these men were asserting their right to accountability by bearing arms. Specifically, they armed themselves with multiple U.S. commercial aircraft. Video footage and flight data recorders have provided documented accountability over their decisions.[481]

On September 11, 2001 (9/11), at 7:59 a.m. EST, four of the men boarded American Airlines Flight 11, leaving Boston for Los Angeles. By 8:15 a.m., air traffic control lost radio contact with the flight. Instead of heading to Los Angeles, decision-makers on Flight 11 descended to 900 feet. Carrying 10,000 gallons of jet fuel, they redirected Flight 11 toward the two towers of New York's World Trade Center. At 8:46 a.m., Flight 11 struck the World Trade Center's north tower at 465 miles per hour. At 9:03 a.m., United Airlines Flight 175 collided with the World Trade Center's south tower at 590 miles per hour. By 10:28 a.m., both towers had collapsed.[482, 483]

New York's World Trade Center had been a monumental symbol of centralized global resources, specifically in the financial markets of Wall Street. Just as on the eves of the first two world wars, global media outlets polarized the world. Middle Eastern officials called the men heroes, martyrs, and freedom fighters.[484] U.S. government officials said they were terrorists, cowards, and murderers.[485] These were the same people U.S. government officials had previously labeled freedom fighters during Operation Cyclone.[486, 487]

2001: NSA Whistleblowers Binney, Drake, and Tamm

In the wake of 9/11, National Security Agency (NSA) Technical Director and whistleblower William Binney internally reported evidence of waste and mismanagement involving a new project called Trailblazer. In an effort to prevent additional terrorist attacks, Trailblazer secretly recorded telephone, email, other communications, and other personal information for every American. Through Trailblazer, NSA officials were violating the Fourth Amendment to the U.S. Constitution by seizing communications without warrants.[488]

Trailblazer cost U.S. taxpayers roughly $1 billion and was purchased from a corporation called Science Applications International (SAI). In a clear conflict of interest, the decision to purchase and implement Trailblazer was made by NSA Deputy Director William Black Jr., the former vice president of SAI. In response, NSA Technical Director and whistleblower Thomas Drake filed a complaint with both the NSA and the Department of Defense (DOD). No investigation was provided, and no one at the NSA or the DOD even responded to Drake's complaint.[489]

Soon after, DOJ attorney and whistleblower Thomas Tamm leaked the details of project Trailblazer to the press. Tamm's house was subsequently raided by FBI agents. Drake then followed up Tamm's efforts by leaking the findings of an NSA Office of the Inspector General (NSA OIG) internal investigation that corroborated Tamm's allegations. The NSA OIG internal investigation concluded the Trailblazer program was operating illegally.[490]

Belatedly, Drake and Binney were also criminally indicted. The retaliatory treatment of Binney, Drake, and Tamm made an unforgettable impression on U.S. government employees such as Edward Snowden. Under U.S. whistleblower laws, government employees are required to whistle blow internally. As Snowden would later describe it, "You have to report wrongdoing to those [who are] most responsible for it."[491] In other words, the defendant is the judge and jury. As such, the whistleblower is effectively put on trial. According to Snowden, "If there hadn't been a Thomas Drake, there couldn't have been an Edward Snowden."[492]

Snowden and other U.S. government whistleblowers learned that internal whistleblowing is ineffective. Internal whistleblowing tips off superiors. By requiring that whistleblowing take place internally, superiors can destroy evidence, build alibies, and manufacture excuses to retaliate against the whistleblower. Labels like "disgruntled," "liar," and "traitor" can then be hung on the whistleblower while superiors maintain the status quo.[493, 494]

From the perspective of coworkers, the whistleblower disappears overnight. From the perspective of taxpayers, self-regulated government agencies perpetually appear to run smoothly.

CHAPTER 20

Self-regulation

Self-accountability vs. Third-party Accountability

Self-regulation is undemocratic because it centralizes accountability power. U.S. government agencies and corporations are permitted to self-regulate. This is a problem because they are not accountable for their actions. The contrast can be seen by examining U.S. accountability mechanisms. The 99% are held accountable using a brutal array of criminal laws. The 1% are held accountable at best by slap-on-the-wrist civil laws and at worst by whistleblowers like Binney, Drake, and Tamm, who are then sitting ducks for retaliation.

In the case of the 99%, criminal laws provide for third-party investigations through a police mechanism. Police collect and control evidence of bad behavior. The discovery process is somewhat objective. If needed, evidence of wrongdoing can be violently seized at gunpoint by a third-party order through a judicial mechanism. Neither the plaintiff nor the defendant controls the evidence. Adjudication is also provided by a third-party through a jury mechanism.

In the case of U.S. government agencies and corporations (the 1%), civil laws allow the accused party to self-investigate. There is no third party investigation. Evidence is controlled by the defendant. When a lawsuit is filed against a government agency or corporation, the defendant is politely asked to produce evidence. The discovery process is subjective. The defendant is conveniently positioned to destroy, cover up, distort, or refuse to produce anything incriminating. In some cases, the defendant self-adjudicates. To use a school analogy, self-regulation allows the 1% to grade their own exams.

2002: Department of Justice Whistleblowers Edmonds and Radack

In 2002 FBI linguist Sibel Edmonds internally reported acts of gross mismanagement to FBI officials. Soon after, Edmonds was fired. Most notably, Edmonds reported information that was omitted from any published 9/11 congressional reports. According to Edmonds, her FBI superiors failed to act in April of 2001 after Edmond's FBI unit was told the following by an informant (a former head of Iranian intelligence):

> "Bin Laden's group is planning a massive terrorist attack in the United States; The order has been issued; They are targeting major cities, big metropolitan cities; they think four or five cities;

> New York City, Chicago, Washington, D.C., and San Francisco; possibly Los Angeles or Las Vegas; They will use airplanes to carry out the attacks; They said that some of the individuals involved in carrying this out are already in the United States."

Under the auspices of protecting an intelligence source (the Iranian), Edmonds was discredited and ultimately fired. A few days after 9/11, the FBI agents handling this informant were ordered by their Special Agent in Charge to remain mute on the subject: "Those conversations never existed; it never happened; this is very sensitive; no one should ever mention a word about this case."[495]

Following a 2004 Department of Justice Office of the Inspector General (DOJ OIG) whistleblower retaliation investigation, the DOJ admitted that many of Edmond's allegations were supported and were the most significant factor in the FBI's decision to fire her.

As a translator for the FBI counterintelligence division, Edmonds also learned how the CIA smuggled heroin from the Middle East to the U.S.: "Drugs were going to Belgium [from Turkey] on NATO planes. After that, they went to the UK, and a lot came to the U.S. via military planes to distribution centers in Chicago and Paterson, New Jersey. Turkish diplomats, who would never be searched, were coming with suitcases of heroin."[496]

About the same time as Edmonds was whistleblowing at the FBI, DOJ attorney and whistleblower Jesselyn Radack internally reported to DOJ officials that DOJ attorneys were suppressing evidence during the prosecution of John Walker Lindh. At the time, photos of Lindh blindfolded and duct-taped naked to a stretcher by U.S. troops were circulating on the Internet. Lindh, a U.S. citizen, had been tortured in Afghanistan by U.S. troops. In response, the DOJ "self-regulated": DOJ officials began covering up evidence supporting Radack's allegations.[497]

Radack was told by her DOJ supervisor to find another job. Soon after, a DOJ official told Radack's new employer she was being criminally investigated and her license to practice law might be revoked. As a result, Radack was fired by her new employer.

It was later discovered Republican Party officials were making all the decisions. Through White House Counsel Alberto Gonzales, Republican Party officials had ordered the Attorney General to illegally withhold evidence. Gonzales later confessed that the White House was calling the shots and that he, as White House counsel, had decided not to turn anything over to Lindh's defense lawyer in the way of documents.[498] Gonzales had a conflict of interest. He understood political party loyalty ensured his job security.

During that same period, Republican Party officials ordered Attorney General John Ashcroft to take three actions:

- force the resignation of Jack Goldsmith, the DOJ attorney who decided torturing war prisoners was unconstitutional[499]
- reverse the Goldsmith legal decision, allowing for the secret and unlawful torturing of war

prisoners[500]

- reverse a prior DOJ legal decision on NSA Project Stellar Wind, allowing for the unlawful surveillance of U.S. citizens[501]

Coincidentally, Ashcroft was then hospitalized with acute pancreatitis. In response, Republican Party officials sent Gonzales to the hospital, where he asked Ashcroft to sign a document reversing his decision regarding Stellar Wind. A half-conscious Ashcroft refused to sign.[502]

After GWB was reelected president, Gonzales's loyalty was rewarded as he was promoted from White House Counsel to U.S. Attorney General. Soon after, GWB ordered Gonzales to dismiss at least twenty-six DOJ attorneys whose past actions had harmed the Republican Party. According to GWB's chief of staff, Kyle Sampson, those attorneys were not "loyal Bushies."[503]

CHAPTER 21

Cover-ups

2003: CIA Whistleblowers Plame and Wilson

In 2003 Joseph C. Wilson, a former U.S. ambassador to Africa and CIA consultant, published the article, "What I Didn't Find in Africa," questioning GWB's motives for invading Iraq. He was also CIA agent Valerie Plame's husband. At the time, Plame was covertly assigned to the Middle East and was privy to whether materials traveling from Africa to Iraq were being used to build nuclear weapons. The article was published following a GWB speech that cited CIA intelligence derived from Plame's work.[504]

Presumably acting at the direction of U.S. government officials, Italy's military intelligence service manufactured the intelligence cited by GWB. The Yellowcake Forgery documents were fake Nigerian government letters, the content of which suggested that Iraq had purchased yellowcake uranium powder, a material used to make weapons of mass destruction.

Italy then propagated the documents to U.S. intelligence agencies. According to GWB, the documents conveniently showed that Iraq was building nuclear weapons, forming the justification for invasion.[505]

In response, Wilson's article claimed that GWB was lying. A few days after Wilson's article was published, the GWB administration, which included Karl Rove and Scooter Libby, retaliated by illegally leaking Plame's top secret CIA employment status to the press, effectively ending her CIA career. Rove initiated a covert propaganda campaign to discredit Wilson as a liar and portray Plame as a below-average CIA agent. Libby was criminally convicted for leaking Plame's identity as a CIA agent, but he was eventually pardoned by Republican president Donald Trump.[506, 507]

Following the invasion of Iraq, Vice President Dick Cheney was asked, "Do you think the American people are prepared for a long, costly, and bloody battle with significant American casualties?" He responded, "Well, I don't think it's likely to unfold that way… I really do believe that we will be greeted as liberators."[508]

In 2004, U.S. weapons inspector David Kay resigned, stating, "I don't think [WMDs] exist [in Iraq]."[509] CIA Director George Tenet then asked Kay not to resign, stating, "If you resign now, it will appear that we don't know what we're doing."[510]

As it turned out, the process of manufacturing a threat had been a few years in the making. As Secretary of State Condoleezza Rice described it in 2000, "We need a common enemy to unite us."[511]

Stated differently, we need a war to unite us. While the GWB administration may have envisioned Iraq as the unifying common enemy, the U.S. 99% learned their enemy was much closer to home.

2004: U.S. Army Whistleblowers Tillman, Greenhouse, and Provance

After 9/11, Pat Tillman turned down a multimillion-dollar National Football League (NFL) contract to join the U.S. Army. After returning from his first tour in Afghanistan, Tillman confessed to a friend, "We don't know who the enemy is."[512]

Tillman then fulfilled his contractual obligation to return to the U.S. Army for a second tour. During that tour in 2004, he identified the enemy. Tillman told other people in his platoon not to reelect GWB as president. Tillman described the war as "so fucking illegal."[513]

Tillman had arranged to meet with liberal antiwar activist Noam Chomsky when he returned home from his second tour. If those communications between Tillman and Chomsky were intercepted by U.S. military surveillance, Republican Party officials would have appreciated the political threat. If a well-known antiwar activist and an army ranger and NFL star spoke out together against the wars, many Americans would have listened. The Tillman-Chomsky "don't reelect Bush" message would have been an incredible threat to Republicans. Tillman and Chomsky never got the chance to meet.

On April 22, 2004, in Spera, Afghanistan, Tillman was shot three times in the forehead. A CIA-trained allied soldier, representing the Afghan Military Forces

(AMF), was also killed.[514] Similar to the JFK-Oswald assassinations, Tillman's assassin may have been assassinated.

According to an Army doctor, it appeared Tillman was shot from approximately ten yards away. But the shots did not come from enemy fire. There was no firefight. No evidence of enemy fire was ever found. The shots came from someone in Tillman's own platoon.[515]

Because the tightly grouped wounds indicated murder rather than the random gunshot wounds characteristic of combat, Army doctors recommended the Army open a criminal investigation.[516] That criminal investigation never took place. Tillman was apparently facing the shooter and may not have considered that person a threat.[517]

Ballistics would have served to identify which weapons fired shots. Tillman's weapon and helmet disappeared. His body armor and uniform were burned on site. Tillman's platoon was also instructed to lie about the circumstances of Tillman's death. The shooting had all the indicators of an assassination and cover-up. Accountability power over Tillman's death was being centralized.[518, 519]

Tillman had the undeniable potential to become a politician. Instead of providing the American people with accountability, the U.S. government rolled out the propaganda machine, spinning Tillman's death as typical of combat. Tillman was awarded a Purple Heart and a Silver Star, as U.S. government officials waited thirty-five days to reveal Tillman was killed by "friendly fire."[520, 521]

According to a Department of Defense inquiry, Army attorneys had deliberately spun Tillman's death as resulting from friendly fire to avoid a criminal investigation. When the U.S. legislative branch tried to check the executive branch by investigating, loyal Republicans like GWB, Vice President Dick Cheney, Secretary of Defense Donald Rumsfeld, and Secretary of State Condoleezza Rice dead-ended the investigation by citing executive privilege.[522, 523]

Halliburton CEO Dick Cheney

In 2004 Army whistleblower Bunny Greenhouse testified before the legislative branch that secret single-bid war contracts were illegally awarded to military-industrial corporations like Halliburton without allowing other corporations to bid. Halliburton, which made well over $20 billion off U.S. taxpayers on the war in Iraq, is the same company that paid Dick Cheney more than $30 million to resign as their CEO and run for vice president in 2000.[524]

Cheney had obvious conflicts of interest. While Cheney and other executive branch decision-makers were shielded from criminal and congressional investigations by the state secrets privilege, Greenhouse was demoted by the Army.[525]

Torture and Abuse of Iraqi Prisoners

In 2004 during the invasion of Iraq, U.S. Army Major General Antonio Taguba conducted an internal investigation into detainee abuse at Abu Ghraib U.S.

military prison in Baghdad. The resulting Taguba Report was then leaked to the public by an anonymous whistleblower.

A second whistleblower, U.S. Army Intelligence Sergeant Samuel Provance, who was listed as an interviewee in the Taguba Report, then disclosed that U.S. military personnel had been encouraged to torture and abuse detainees. According to the Taguba Report, abuse of detainees included many violent or humiliating actions:

> "(1) punching, slapping, and kicking detainees; jumping on their naked feet
> (2) videotaping and photographing naked male and female detainees
> (3) forcibly arranging detainees in various sexually explicit positions for photographing
> (4) forcing detainees to remove their clothing and keeping them naked for several days at a time
> (5) forcing naked male detainees to wear women's underwear
> (6) forcing groups of male detainees to masturbate themselves while being photographed and videotaped
> (7) arranging naked male detainees in a pile and then jumping on them
> (8) positioning a naked detainee on an MRE box, with a sandbag on his head, and attaching wires to his fingers, toes, and penis to simulate electric torture
> (9) writing 'I am a Rapest' [sic] on the leg of a detainee alleged to have forcibly raped a

fifteen-year-old fellow detainee, and then photographing him naked
(10) placing a dog chain or strap around a naked detainee's neck and having a female soldier pose for a picture"[526]

In response to the Taguba Report and other reports on torture, the legislative branch opened a congressional investigation. The investigation determined that Attorney General Gonzales had secretly authorized the covert use of torture techniques such as mock executions, walling, cramped confinement, up to 180 hours of sleep deprivation, insects placed in a confinement box, and waterboarding.

In addition, the CIA had engaged in the following egregious tactics: "Taking power drills to the heads of captured men; making them stand with their arms stretched above their heads for days at a time; leaving at least one of them naked until he froze to death; waterboarding them to the point of catatonia as bubbles rose from their open mouths; and inserting pureed food into their rectums while claiming it was necessary for delivering nutrients."[527]

Provance resisted the Army's cover-up and was placed in a career hold, losing his security clearances. He was then threatened with ten years in a military prison, demoted, and discharged.[528, 529, 530]

2005: NSA Whistleblower Tice

In 2005 NSA analyst and whistleblower Russell Tice publicly requested that the U.S. legislative branch pass stronger protections for whistleblowers. Tice reported that communications channels for journalists and news organizations were secretly being recorded by the NSA twenty-four hours a day, seven days a week. By tracking every communication ever sent to or from a journalist, whistleblowers could be stopped cold.

Tice also reported that NSA programs were secretly and deliberately given both military and intelligence oversight status such that each status could be denied in view of the other status, leaving both programs oversight-free. Tice described it the following way:

> "I learned the hard way [that] you cannot trust any of the internal supposed mechanisms that are there for oversight. The chain of command, the IG's office (even at the DOD IG, I found) was basically trying to put a knife in my back. The Whistleblower Protection Act does not apply to the intelligence community. They're exempt from it. And most people in the intelligence community, they don't realize that. So you can't even go to the Office of Special Counsel because they're exempt from that too, and the merit system protection board. So even if you use the whistleblower... intelligence community Whistleblower Protection Act, the only thing that gives you is the right to go to Congress. It doesn't... it doesn't have any teeth there to protect you against retribution from the agency that you're reporting abuse on."[531]

Tice was eventually fired by the NSA.[532, 533]

2006: Department of the Interior Whistleblowers Maxwell, Little, and Morris

In 2006 Department of the Interior (DOI) whistleblowers Bobby Maxwell, Randall Little, and Lan Morris reported to DOI officials that their supervisors refused to collect on debts owed to the government by oil and gas companies. A subsequent investigation by the DOI Office of the Inspector General (DOI OIG) supported the allegations, concluding that the DOI had lost up to $10 billion in government revenue due to mismanagement.

DOI officials had conflicts of interest. Internal investigations determined that several DOI officials frequently consumed alcohol, used cocaine and marijuana, and had sexual relationships with the oil and gas company representatives they met at industry functions. They had effectively been seduced by the oil and gas companies they were responsible for regulating.

During that period, DOI officials were described as having stonewalled the U.S. legislative branch regarding concerns of a "catastrophic disaster in the Gulf of Mexico." DOI Inspector General Earl Devaney confessed: "Short of a crime, anything goes at the highest levels of the [DOI]."[534]

In April of 2010, national environmental security was compromised when a British Petroleum oil rig, which had been informally exempted from DOI safety regulations by DOI officials, ruptured in the Gulf of Mexico. Despite eleven deaths and catastrophic environmental harm to the gulf, British Petroleum executives were never prosecuted. While DOI officials were shielded from civil and criminal liability, Maxwell, Little, and Morris were demoted and threatened with criminal prosecution.[535, 536, 537]

CHAPTER 22

Fraud

2007: Lack of Corporate Accountability

Corporations are a legal fiction that allow the 1% to minimize accountability over themselves. Like hiding behind a mask, incorporating allows accountability information, such as identity information, to be hidden from the 99%.

In 2007 a Drug Enforcement Administration (DEA) official received an email from a man named Enrique Prado that explained his employer, Blackwater USA Incorporated, could "do everything" and that "deniability is built in and should be a big plus." Prado, as it turns out, was a former CIA agent. According to Blackwater's founder, Blackwater was "trying to do for the national security apparatus what FedEx did for the Postal Service."[538]

In other words, Blackwater wanted the government to outsource its covert actions, such as assassinations. This way, government officials had deniability. They could shield themselves from liability and fraudulently claim no knowledge of covert corporate operations. Prado's background was subsequently investigated by author Evan Wright.

According to Wright, Prado was a Cuban-born American. After Castro took power in Cuba in 1959, Prado's family was relocated to Miami. Once in the U.S., Prado became a hitman for the mafia. After being hired by the CIA, Prado maintained his mafia cover while carrying out assassinations for the CIA. Following the assassinations of JFK, Robert Kennedy, and MLK, Prado coincidentally experienced a meteoric ascension through the CIA's ranks. By the 1990s, Prado was a senior official in the CIA's Bin Laden unit and eventually became chief of operations for the CIA's Counterterrorism Center. After 9/11, Prado left the CIA to work for Blackwater's "Targeted Assassination Unit."[539, 540]

When the 1% creates targeted assassination units using corporations, it is an act of national security. When the 99% do it, it is an act of terrorism. While the identities of the 99% are rigorously tracked at all times, the 1% has quite a different experience. Like changing ski masks, the 1% can change identities at any time by reincorporating as a new strawman.

In 2009, after receiving bad press for killing seventeen Iraqi civilians, Blackwater changed its name to Xe Services.[541] In 2010 Xe Services received a $250 million contract from the CIA.[542] In 2011 private investors acquired Xe Services. The private investors subsequently changed the name of the corporation from Xe Services to Academi.[543] In 2014 Academi was acquired by Constellis Holdings.[544] In 2015 Academi supplied mercenaries to fight Houthis in Yemen.[545]

In 2016, Academi earned over $500 million on "counter-narcotics" contracts in Afghanistan.[546] According to U.S. Marine Sergeant Matthew Hoh, NATO forces occupying Afghanistan were "more or less guarding poppy fields and poppy production under the guise of counterinsurgency. The logic was 'we don't want to take away the livelihoods of the people.' But really, what we were doing at that point was protecting the wealth of our friends in power in Afghanistan."[547] At the same time, the number of heroin-related overdose deaths quadrupled in the U.S. between 2002 and 2013.[548]

Other corporations serving as the invisible arms of the military-industrial complex include HBGary Incorporated and Strategic Forecasting Incorporated (Stratfor). Stratfor actually markets itself as a "shadow CIA." HBGary has consulted Bank of America regarding strategies for launching cyberattacks on the WikiLeaks website. Likewise, Stratfor has provided the U.S. Department of Homeland Security with analytics used to break up the Occupy Wall Street and Black Lives Matter movements.[549]

Both HBGary and Stratfor have helped hack the personal data of journalists who publish information critical of corporations. Clients of HBGary and Stratfor include Dow Chemical, Coca-Cola, Goldman Sachs, Northrop Grumman, Lockheed Martin, and Raytheon. As long as the 1% has their backs, nobody can hold corporations like HBGary, Stratfor, Blackwater, Xe Services, or Academi accountable.[550]

2008: Overcoming U.S. Election Rigging

Pursuant to Article I of the U.S. Constitution, each state governs its own election mechanisms. Less than 1% of the U.S. population accounts for the votes. During U.S. presidential elections, vote totals for each state are centralized and accounted for in a variety of corporate vote-counting computers and mass media outlets. To understand the fraud risks associated with centralized election mechanisms, consider the 2000 U.S. presidential election and every U.S. presidential election thereafter.

Diebold Centralized Vote-counting Computers

During the night of the 2000 U.S. presidential election, at approximately 10:15 p.m. EST, Florida experienced an unaccounted-for vote flip. According to one investigation, Deborah Tannenbaum, a poll worker in Volusia, Florida "called the county elections department and learned that Al Gore was leading GWB 83,000 votes to 62,000. But when she checked the county's website for an update half an hour later, she found a startling development: Gore's count had dropped by 16,000 votes."[551]

Democratic presidential candidate Al Gore magically lost 16,000 votes. Volusia had been using vote counting computers supplied by Diebold Election Systems. A few years later, internal emails from Diebold were published online. In one of those emails, Diebold employee Lana Hines stated the following: "I need some answers! Our department is being audited by the County. I have been waiting for someone to give me an explanation as to why Precinct 216 gave Al Gore a minus 16,022 when [the count] was uploaded. Will someone please explain this so that I have the information to give the auditor instead of standing here 'looking dumb.'"[552]

Diebold never issued a public explanation, and no accountability was provided by law enforcement agencies. After a tie was declared in the Florida 2000 presidential election results, the Supreme Court, which was controlled by the Republican Party (with five Republican justices and four Democratic justices), decided the Republican presidential candidate, GWB, was the winner of Florida (the ruling split along party lines).

As a result, GWB obtained Florida's twenty-five electoral votes. This gave him 271 electoral votes compared to Gore's 266, clinching the 2000 presidential election for GWB, the son of former CIA director George H.W. Bush and the brother of Florida Governor Jeb Bush.[553]

Successfully rigging the election in only one state can change the outcome of an entire election. Had GWB lost Florida, Gore would have captured 291 electoral votes, and GWB would have only secured 246. But for Florida in 2000, Gore would have become president.

Sequoia Voting Systems

During the Florida 2000 presidential election, paper ballots had been counted using centralized electronic vote-counting machines. After the election, seven whistleblowers who were employees of Sequoia Voting Systems in 1999 testified that paper ballots in Florida were deliberately manufactured using inferior paper stock. The paper ballots had been excluded from quality assurance testing and deliberately left misaligned. Misalignments would create problems such as hanging chads. The goal was to discredit paper ballots and make both the ballot process and the vote counting process entirely electronic.[554, 555]

As a result of the hanging chads problem in the Florida 2000 presidential election, the U.S. legislative branch passed the Help America Vote Act, which gave roughly $4 billion to the electronic voting industry. In addition, all fifty states were required to make their vote counting systems electronically centralized by 2004.[556, 557]

Election Rigging in Georgia

In 2002 whistleblower Chris Hood, a Diebold contractor, claimed that Diebold was covertly installing unauthorized software onto electronic voting machines. Hood alleged that Diebold had complete statewide control over the Georgia 2002 election results.[558]

That same year, whistleblower Rob Behler, also a Diebold contractor, claimed that Diebold was installing unauthorized software without the consent of Georgia election officials onto Georgia's electronic voting machines. Georgia's governor and senator both lost in landslides after widely being expected to win, resulting in Georgia's first Republican governor since 1872.[559, 560]

Election Rigging in California

In 2003 whistleblower Stephen Heller accused Diebold of installing unauthorized software onto electronic voting machines in Alameda County, California.[561]

In 2004 California Secretary of State Kevin Shelley, a Democrat, decertified all Diebold electronic voting machines and recommended criminal prosecution of the company. The California Governor's Office, under Republican Arnold Schwarzenegger, then blocked the proposed criminal prosecution of Diebold.[562]

In 2006, California Secretary of State Bruce McPherson, a Republican, recertified Diebold for California elections.[563]

Election Rigging in Ohio

During the 2004 U.S. presidential election, Diebold and its sister company Election Systems & Software (ES&S) had centralized control over a majority of U.S. electronic vote-counting machines. Whistleblower Bob Magnan, an IT specialist for the state of Ohio, was working the night of the election. Magnan was running the centralized computers responsible for counting votes for the entire state. Unexpectedly at 9:00 p.m. EST, Magnan's boss, Ohio Republican Governor Ken Blackwell, sent Magnan home so that corporate contractors could take over.[564, 565]

Sometime thereafter, the Ohio vote-counting computers crashed. Control over Ohio's vote-counting system was switched from Ohio government computers to computers in Tennessee, the headquarters of

SmarTech Corporation. A 6% vote flip then occurred, leading to the reelection of Governor Blackwell and a win in Ohio for GWB.

The Ohio win provided GWB with twenty electoral votes in his reelection effort. This gave him 286 electoral votes compared to Democratic Party candidate John Kerry's 251, clinching GWB's reelection.[566] Had GWB lost Ohio, Kerry would have finished with 271 electoral votes, and GWB would have finished with 266. But for Ohio in 2004, Kerry would have become president.

The computers for the 2004 Ohio vote count had been set up through SmarTech by a man named Michael Connell. Connell was an IT consultant to Blackwell, GWB, Karl Rove, and others in the Republican Party. In September 2008, Connell was subpoenaed to testify in a lawsuit alleging election rigging during the 2004 presidential election.

During litigation, the plaintiff's attorneys wrote the U.S. Department of Justice requesting witness protection for Connell, alleging the following: "We have been confidentially informed by a source we believe to be credible that Karl Rove has threatened Michael Connell, a principal witness we have identified in our King Lincoln case in federal court in Columbus, Ohio, that if he does not agree to 'take the fall' for election fraud in Ohio, his wife Heather will be prosecuted for supposed lobby law violations."[567, 568]

During pretrial depositions, Connell claimed that on election night 2004, a website called "gwb43" connected Rove to SmarTech computers in Tennessee. In December 2008, only weeks before the trial, Connell was killed when his single-engine plane mysteriously crashed while flying from Washington, D.C., to his home in Ohio.[569]

Hacking the U.S. Election Process

Following the 2004 U.S. presidential election, Princeton University professors Andrew Appel and Ed Felten, both computer scientists, wanted to see how easily the U.S. electronic vote-counting process could be hacked. A whistleblower at Diebold provided Appel and Felten a Diebold machine. The professors were able to unlock the computer on the machine using a standard key common to locks used on filing cabinets and hotel minibars. Using the default administrative password "1111," they also installed vote-flipping software that replicated the vote-flipping process allegedly used during the 2004 U.S. presidential election.[570]

Along the same lines, Sequoia Voting Systems had been marketing their Advantage electronic voting machines as tamperproof. The professors went online and purchased machines manufactured by Sequoia. Sequoia's legal team subsequently threatened both professors with legal action should Princeton conduct studies of Sequoia's machines. Sequoia attorneys insisted that their software was a proprietary trade secret protected by intellectual property laws and that it could not be lawfully audited or accounted for.[571]

On the night of the 2012 U.S. presidential election at approximately 11:25 p.m. EST, Democratic presidential candidate Barack Obama was declared the winner of Ohio by state officials. Fox News then announced that Obama had won the entire election. One of the Fox News commentators, Karl Rove, was incredulous, as if he was awaiting another Ohio vote flip. Rove had earned the nickname "Turd Blossom" from GWB while serving as campaign advisor during the 2000 and 2004 presidential elections. It was a Texas reference to the flowers that bloom out of cattle poop, implying that Rove could turn a turd into a U.S. president. Had a vote flip occurred that night in Ohio, Republican presidential candidate Mitt Romney would have become president instead of Obama.[572]

Two weeks before the Fox News broadcast, a group called Anonymous posted a video that warned Rove not to rig the election. In the days after the Obama victory, a group linked to Anonymous published a statement claiming that they had created password protected firewalls around vote counting computers in several states, including Ohio, to block Rove's Republican Party operatives. According to Anonymous, Rove's operatives had hacked centralized vote-counting databases during previous elections, allowing Republicans to secretly flip vote counts. By blocking remote access to centralized government databases, vote rigging was allegedly minimized during the 2008 and 2012 presidential elections, both of which Obama won.[573]

2009: Accountability over Life-essential Resources

Under California law, a landowner can take as much water out of the ground as they want, even if that water is pumped from hundreds of miles away and from under another landowner's property. To exploit this law, a small group of investors arranged a secret meeting with California government officials in 1994. The meeting, which took place in Monterey, California, resulted in amendments to an existing contract between incorporated farmers and the State of California.[574]

In short, the Monterey Amendments secretly transferred a substantial portion of California's underground water supply from the people of California to a water storage corporation. As of 2018, that corporation was owned by Roll Corporation International. Roll is the billion-dollar parent company of products like Fiji Water, POM Wonderful, Wonderful Pistachios, and Wonderful Almonds. Under the Monterey Amendments, California taxpayers have to buy back the water in times of drought.[575]

During the drought of 2009, California taxpayers had to do just that. The Monterey Amendments became public knowledge after an anonymous whistleblower contacted a citizens' rights group. Rather than criminally investigating the government officials who made the agreement, politicians are working to gradually unroll the Monterey Amendments over several decades. Coincidentally, these politicians have been the recipients of campaign donations from Roll Corporation.[576]

San Bernardino, California

For decades, a permit from the U.S. Forest Service has allowed Nestlé, a Swiss corporation, to pump and bottle water out of the publicly owned San Bernardino National Forest. After that permit expired in 1988, the head of the U.S. Forest Service for the San Bernardino National Forest, Gene Zimmerman, insisted that the permit remain valid until revoked. After retiring in 2005, Zimmerman became a paid consultant for Nestlé. He is also a board member for a Nestlé-funded non-profit. When confronted, Zimmerman stated, "I don't see any conflict."[577]

Nestlé is pumping approximately 27 million gallons per year for its Arrowhead brand from the San Bernardino National Forest. At the same time, the People of California are in drought lockdown. While creek beds in the San Bernardino National Forest dry up, Nestlé continues to quietly pay approximately $600 per year for the permit and nothing for the water.[578]

McCloud, California

In 2003 Nestlé secretly negotiated an agreement with McCloud politicians to build a water-bottling plant. In one short-notice public hearing, Nestlé provided local residents with a slide presentation touting the company's environmental accomplishments. Moments later, without debate, city politicians voted in favor of the agreement. Under it, Nestlé would own McCloud's water supply for the next 100 years. In addition, the agreement allowed the following actions:

(1) No Environmental Impact Statement (EIS) was required or produced.

(2) More than 200 tanker trucks per day would travel through McCloud.
(3) The City of McCloud would pay for the infrastructure and environmental impact.
(4) Nestlé would pay less than 10% of the national market value for the water.
(5) The price Nestlé paid would be fixed.
(6) Nestlé would have "first rights" to the water over the people of McCloud.[579, 580]

In times of drought, the people of McCloud have to buy back their own water.[581] Nestlé has attempted to forge similar agreements in other cities:

- Chafee County, Colorado
- Mecosta County, Michigan
- Cascade Locks, Oregon
- Fryeburg, Maine
- Enumclaw, Washington
- Kunkletown, Pennsylvania
- Wacissa, Florida
- Aberfoyle, Ontario[582]

Coca-Cola and Pepsi are also centralizing water resources and making billions in profits. Coca-Cola's Dasani brand and Pepsi's Aquafina brand, both of which are also bottled in California, contain "purified municipal water." While testing results for U.S. tap water are published by the EPA, the bottled water industry self-regulates.

Testing results for bottled water are kept secret by the Food and Drug Administration (FDA). Private studies have shown little difference in quality between California tap water and bottled water, yet the corporate spin is clear: bottled water is cleaner, safer, and purer than tap water. Under the current system, anyone in California can use a water hose to fill a plastic bottle and sell it as clean, safe, and pure.[583]

If the American people are buying their own water when they buy bottled water, then the only value obtained is trillions of plastic bottles. And therein lies a double whammy: plastic water bottles are made out of refined crude oil. Oil companies make money off bottled water too.

Flint, Michigan

In 2010 venture capitalist Rick Snyder was elected Governor of Michigan. In 2014 Snyder decided to change the city of Flint's drinking water supply from Detroit Water to Karegnondi Water Authority (KWA), a municipal corporation. In an attempt to keep Flint connected to Detroit Water, Detroit Water officials offered to reduce their price by 50%. The offer was refused.

Until the KWA pipeline was built, the Flint water supply would be switched to the Flint River. After the switch, odorous, discolored water began pouring from taps.[584, 585] Residents developed rashes and stomach problems. At least twelve people died from an outbreak of Legionnaires' disease.[586] Both lead and fecal coliform bacterium were detected in the Flint River tap water.

In 2016, Snyder took home three gallons of Flint tap water and promised to drink it for thirty days. One week later, Snyder left for Europe.[587]

CHAPTER 23

Whistleblowing

2010: Chelsea Manning

In 2010 whistleblowing became trendy, thanks to a new social mechanism called WikiLeaks. WikiLeaks was an Internet website created to decentralize accountability information. Once information was uploaded to WikiLeaks by a whistleblower, the information would be published, as well as decentralized to databases located all over the world.[588]

One such whistleblower was U.S. Army intelligence analyst Chelsea Manning. Manning was ignored after reporting evidence of human rights violations to superiors. According to chat log entries made before she was arrested, the incident that affected Manning most was when she discovered that fifteen so-called insurgents were actually Iraqi civilians imprisoned for nonviolently disseminating government accountability information. They were publishing scholarly critiques of corruption by the Iraqi prime minister.[589]

These Iraqi civilians were essentially asserting their right to accountability.

Manning reported her findings to her commanding officer, who told Manning he didn't want to hear any of it.[590] Manning's commanding officer had a conflict of interest. Like many U.S. government supervisors, he decided that remaining silent—and thereby securing his paycheck—would be a better option than holding superiors accountable.

According to chat logs documented before her arrest, Manning came to a critical realization: "I was actively involved in something that I was completely against." She added, "I want people to see the truth, regardless of who they are, because without information, you cannot make informed decisions as a public." [591] Manning successfully asserted her right to accountability by providing WikiLeaks evidence of, among other things, human rights violations.

The accused party, the U.S. Army, subsequently asserted centralized accountability power and imprisoned Manning in its own facilities while self-policing the matter. No constitutional judicial process was ever provided. Pursuant to a military process, as opposed to Fifth Amendment due process, Manning was held by the defendant for more than 800 days without a trial. A U.S. Army judge eventually sentenced Manning to thirty-five years in prison.[592, 593]

2011: Sexual Assault Whistleblowers

In 2011, seventeen U.S. military veterans filed *Cioca v. Rumsfeld*. The lawsuit alleged a culture within the

U.S. military of tolerating abuses of power by senior officials seeking sexual gratification from subordinates. The case was dismissed. As described by Pentagon officials, allowing subordinates to hold superiors accountable "undermines command authority."[594, 595]

According to one study, approximately 25% of women in the U.S. military have been sexually assaulted, and roughly 80% have been sexually harassed.[596]

President Donald Trump articulated the problem when he was recorded while saying, "I just start kissing them. It's like a magnet. Just kiss. I don't even wait. And when you're a star, they let you do it. You can do anything. Grab 'em by the pussy, you can do anything."[597]

As of 2020 Trump has been accused of sexual misconduct by no fewer than twenty-five women, but no prosecution had occurred.[598] The reason for this is that the 1% is above the law. As Trump's admission implies, abuse of power is rarely an isolated incident. It is not white entitlement or male entitlement; it is power entitlement. It is a life pattern.

Many careers have been sacrificed by people who said "no" to the 1%. That no comes at a heavy social, economic, and emotional price. Other victims remain silent. Some victims have reported that in the moment, they lost their physical ability to speak.[599, 600] Trump misinterpreted this phenomenon as women letting him "do anything."

The Catholic Church and the Church of Scientology

For the last half century, documented reports of child molestation filed against approximately 3,000 priests were covered up by Catholic Church officials. Once reported, accused priests were simply reassigned to other churches where sexual abuse continued. Catholic Church officials have since acknowledged a culture in which sexual abuse by priests is covered up. Apparently, protecting the reputation of the Catholic Church, or the Catholic 1%, has been a higher priority than protecting children.[601]

Like the Catholic Church, the Church of Scientology 1% appears to prioritize secrecy over protecting its members. Some of Scientology's youngest members, who are often born into the church, are pressured into signing billion-year contracts of employment. They then work, for less than $100 per week, on church-owned cruise ships. The ships sail in international waters, outside the jurisdiction of U.S. law enforcement. Scientology officials on these ships have been accused of physical and sexual assault. When victims attempt to hold these officials accountable, the response from the church is shaming, surveillance, and litigation.[602, 603, 604]

Much as the Scientology 1% and the Catholic 1% defend secrecy in the holy name of the church, the U.S. 1% defends secrecy in the holy name of national security. In the absence of accountability, secrecy allows the 1% in any institution to perpetually abuse power.

2012: The Office of Personnel Management Inspector General

In 2012 Democratic Party member Katherine Archuleta served as director of the Obama 2012 Presidential reelection campaign. After his reelection, Obama rewarded her with the option of becoming head of the Office of Personnel Management (OPM). Archuleta was chosen over significantly more qualified candidates. Instead of hiring the best person for the job, political party loyalty appeared to be rewarded.

In March of 2015, the OPM's inspector general blew the whistle on "persistent deficiencies in OPM's information system security program... [including] incomplete security authorization packages, weaknesses in testing of information security controls, and inaccurate Plans of Action and Milestones."[605]

In April 2015, a massive data breach was discovered at OPM. From approximately 2014 to 2015, Chinese covert agents were hacking OPM and downloading the personal data of approximately 21 million U.S. government employees, former employees, and applicants. The database consisted of U.S. Standard Form-86 (SF-86) job applications, which included residential addresses, telephone numbers, social security numbers, photographs, and fingerprints. The breach presumably exposed undercover agents and intelligence officers.

As FBI Director James Comey described it, "It is a very big deal from a national security perspective and from a counterintelligence perspective. It's a treasure trove of information about everybody who has worked for, tried to work for, or works for the United States government."[606]

Although Archuleta resigned three months later, the damage was done. By appointing government officials based on political party loyalty, government mechanisms become inefficient and, in some cases, incompetent.[607]

2013: Edward Snowden

In 2013, NSA analyst and whistleblower Edward Snowden fled to Hong Kong and leaked thousands of classified documents to international journalists. The documents revealed evidence of unconstitutional warrantless surveillance of the American people by the U.S. executive branch.[608] During a connecting flight from Hong Kong to South America, the U.S. government withdrew Snowden's passport, stranding him in Russia.[609]

According to one of his NSA coworkers, Snowden was "a genius among geniuses."[610]

Snowden claims that his breaking point was "seeing the director of National Intelligence, James Clapper, directly lie under oath to Congress."[611]

Charged with theft of government property and espionage under the Espionage Act, Snowden cannot return to the U.S. without facing criminal prosecution. Violation of the Espionage Act is punishable by death or up to thirty years in prison.[612]

Snowden and other whistleblowers have been the only sources of accountability information involving several surveillance programs in which the U.S. executive branch has warrantlessly spied on the American people:

- *Echelon:* The Echelon program records all information transmitted between orbiting telecommunications satellites. Echelon can access the telephone conversations of Americans and non-Americans. In the 1990s President Bill Clinton leveraged Echelon for campaign donations. The NSA, through Echelon, discovered that a European corporation was bribing the Saudi Arabian government over an aircraft contract. As a result, Boeing was able to secure the $6 billion contract. Similarly, the CIA, through Echelon, discovered that a French corporation was bribing the Brazilian government over a rainforest data contract. As a result, Raytheon was able to take over the $1.3 billion contract. Both Boeing and Raytheon have since donated millions to Clinton presidential campaigns.[613, 614]
- *Mainway:* The Mainway program records all information for telephone calls made through AT&T, Verizon, and other subsidiary communications corporations. The information recorded includes caller name, caller number, receiver name, receiver number, date, time, and duration.[615]
- *Stellar Wind:* The Stellar Wind program records all Internet communications, including emails, telephone conversations, and financial transactions.[616]

- *Dishfire:* The Dishfire program records all SMS data, such as text messages. Globally, Dishfire records more than 200 million text messages and more than 1 million border crossings per day.[617]
- *Prism:* The Prism program records stored Internet communications, such as postings, photos, and videos, from "partner" corporations. Partners have included Microsoft, Yahoo, Google, Facebook, and Apple.[618]
- *XKeyscore:* The XKeyscore program records all email conversations, social media, Internet browsing, geo-positional data, and other information electronically communicated by targeted individuals, whether American or not. As Glenn Greenwald described it, NSA analysts using XKeyscore can "listen to whatever emails they want, whatever telephone calls, browsing histories, Microsoft Word documents. And it's all done with no need to go to a court, with no need to even get supervisor approval on the part of the analyst."[619]
- *CherryBlossom:* With the secret help of the Stanford Research Institute (SRI), the CIA created CherryBlossom software for use on household wireless routers. The CIA can then remotely monitor email and other Internet traffic that passes from a target device through the compromised router to the Internet.[620]

The CIA has also deployed WeepingAngel, After-Midnight, Athena, DarkMatter, Grasshopper, Pandemic, and Hive. Each of these programs is illegal because no search warrant is obtained by the executive branch.[621, 622, 623]

CHAPTER 24

Retaliation

The King Can Do No Wrong

Roughly seven years before Snowden's revelations, AT&T technician and whistleblower Mark Klein revealed that the U.S. 1% had secret rooms in AT&T buildings around the world where it could record telephone conversations without a search warrant. As part of a checks-and-balances process, the U.S. legislative branch opened a congressional investigation. The investigation was instantly dead-ended by the executive branch, citing the state secrets privilege. There was no check, let alone a balance.[624]

Under U.S. law, the U.S. government has centralized accountability power through the doctrine of sovereign immunity. In other words, the U.S. government cannot be sued. Sovereign immunity is derived from the British common law doctrine that "the king can do no wrong." U.S. government officials can be sued as individuals, but the U.S. government cannot be sued without its own consent. On top of that, individual government officials can then assert state secrets privilege or executive privilege.

The state secrets privilege also derives from British common law. It allows U.S. executive branch officials to avoid accountability by claiming that their testimony would harm national security. All that is required is an affidavit from the U.S. government stating that "producing the evidence, cited by the plaintiff, might harm national security."

Simply put, the judicial and legislative branches are unable to hold the executive branch accountable. As a result, whistleblowers like Snowden cannot account for their accusations without stealing the evidence, in violation of espionage and other criminal laws. This combination of sovereignties, privileges, immunities, and criminal laws is a very powerful collection of mechanisms for retaliating, discouraging future whistleblowing, and generally maintaining centralized power.

Above the Law: Democratic Presidential Candidate Hillary Clinton

When asked about Snowden, then U.S. Secretary of State Hillary Clinton responded thusly:

> "He could have been a [statutory] whistleblower. He could have gotten all the protections of being a [statutory] whistleblower. He could have raised all the issues that he has raised, and I think there would have been a positive response to that. In addition, he stole very

> important information that has unfortunately fallen into a lot of the wrong hands. So, I don't think he should be brought back home without facing the music."[625]

By saying Snowden could have received all the "protections" of being a statutory whistleblower, perhaps Hillary Clinton was referring to hostility, stigmatization, demotion, forced resignation, termination, loss of security clearances, unemployment, student loan default, foreclosure, bankruptcy, loss of credit, loss of hirability, humiliation, and criminal prosecution.

Apparently, when Democrat or Republican officials steal classified documents, they're above the law. In 2015 Hillary Clinton herself avoided facing the music from her actions as secretary of state. According to the FBI, Hillary Clinton sent 110 personal emails containing classified information, of which sixty-five were classified secret and twenty-two were classified top secret. For any other American, the punishment would have been up to 325 years in prison. For the face of the Democratic Party and the candidate Democrats were certain, at the time, would win the 2016 presidential election, crimes get swept under the rug.[626] It was a clear abuse of power by a Democratic presidential administration.

Coincidentally, during that period, former CIA Director and Republican David Petraeus also avoided facing the music. According to the FBI, Petraeus had criminally provided classified information to his mistress and then lied about it to investigators. As of

2020, neither Hillary Clinton nor David Petraeus had spent a day in prison.[627] Snowden, on the other hand, is effectively imprisoned in Russia and may at some point become a bargaining chip for Russian officials.[628]

2014: Russian Whistleblower Retaliation

In 2014 Mikhail Lesin moved to Beverly Hills, California. Despite living most his life on the salary of a Russian government employee, Lesin purchased over $28 million in U.S. real estate. As a government official in Russia, Lesin's resources far exceeded his paycheck. Two decades earlier, Russian President Putin made Lesin the minister of the press—the office responsible for Russian propaganda.

Between 1999 and 2014, Lesin centralized Russian communication power by seizing media outlets. This earned Lesin the nickname "the Bulldozer." Lesin also created the global television network Russia Today (RT). According to Lesin, the goal of RT was "to establish a news channel that would counter CNN, and BBC, with a Moscow spin."[629]

Eventually, Lesin's U.S. shopping spree caught the attention of U.S. law enforcement. The night before a planned meeting with DOJ investigators, Lesin was found dead in a Washington, D.C., hotel room. According to Russian-controlled media, Lesin had died of a heart attack. According to U.S.-controlled media, Lesin was drunk, fell, and fatally hit his head. According to an FBI whistleblower, Lesin was beaten to death.[630]

The prevailing theory is the former head of Russian propaganda was about to reveal Russian secrets to the DOJ in exchange for avoiding U.S. criminal prosecution for financial crimes. The DOJ, however, had paid for Lesin's D.C. hotel room and had reserved the room in Lesin's name. Russian assassins would have had no problem finding him there. Perhaps the Russian 1% is lying to cover up an assassination. Perhaps the U.S. 1% is lying to cover up a DOJ blunder. Either way, two governments abused their power, resulting in nearly zero accountability.[631]

The assassination is a reminder of the global reach of covert agencies. Agents of the FSB, CIA, MI6, and others have the ability to appear and disappear with minimal accountability. When whistleblower and former FSB agent Alexander Litvinenko revealed that the FSB was in fact responsible for the 1999 Russian apartment bombings, the siloviki had Litvinenko secretly poisoned to death in London using a radioactive weapon involving Polonium-210.[632]

Similarly, when Russian auditor Sergei Magnitsky blew the whistle on fraud by Russian government officials, including Putin, he was arrested and died in police custody.[633] Eight other people involved in the Magnitsky case have since died under mysterious circumstances.[634]

2015: Accounting for Russian Assassinations

As documented by the International Consortium of Investigative Journalists, the global 1% still launders and hides money using shell corporations and offshore bank accounts. Like inverted welfare, these social mechanisms allow a global 1% to avoid paying taxes and pass the money through to anonymous or mobile investments, such as real estate and yachts.[635] While they remain off the radar from average law enforcement agencies, the global 1% is not immune to violence by covert agencies.

Since becoming Russia's president, Putin and the siloviki have covertly assassinated numerous oligarchs in pursuit of centralized power. This would be tantamount to the CIA killing Elon Musk, Mark Zuckerberg, Jeff Bezos, Mike Bloomberg, Peter Murdoch, and Bill Gates. The deaths of uncooperative Russian oligarchs are typically staged as suicides.

Following the collapse of Soviet Russia, Boris Beretovsky was the oligarch who controlled Russia's main television network, Channel One. He also owned part of the oil company Sibneft and was a politician. With an estimated wealth of $3 billion, Beretovsky was found dead in his home in 2013. When first responders entered the home, radiation meters were set off. There was evidence of radioactive materials and strangulation, as well as fractured ribs and a wound on the back of Beretovsky's head.[636, 637]

Several additional Russian oligarchs have since died under suspicious circumstances, including Sergey Protosenya (energy sector, found hung to death), Vladislav Avaev (energy sector, found shot to death), Vasily Melnikov (healthcare sector, found stabbed to death), Mikhail Watford (energy sector, found hung to death), Alexander Tyulyakov (energy sector, found hung to death), and Leonid Shulman (energy sector, found stabbed to death).[638, 639]

2016: Assassinations vs. Due Process

Accountability is an evidence collection process. For example, due process under the Fifth and Sixth Amendments to the U.S. Constitution provides for a jury trial. In the case of capital punishment (the death penalty), the decision to kill a defendant is democratically decentralized to jurors. The jury represents a cross-section of the community. In the case of assassinations, the decision to kill is centralized in one member of the community, such as a government official like Putin. Assassinations are therefore undemocratic.

Over recent years, several anonymous U.S. government whistleblowers have revealed a secret U.S. drone assassination program. The Joint Special Operations Command (JSOC) is a covert arm of the U.S. military secretly operating in more than seventy countries, without the consent of the U.S. legislative branch. Investigative journalists estimate that more than 300 civilians have been collaterally killed by negligent JSOC drone assassinations.[640]

According to one JSOC drone operator, assassination targets are followed remotely by a U.S.-based NSA team called GEOCELL. Working with the CIA, FBI, and military, GEOCELL compiles a list of cellular telephone numbers that allegedly have had contact with terrorists. Using global positioning software, GEOCELL then tracks the position of the Subscriber Identity Module (SIM) card corresponding to that cellular telephone number. The SIM card is then hit with a drone-launched air-to-surface missile. The person(s) in the vicinity of the SIM card may or may not be suspected of terrorism.[641]

According to Air Force analyst and whistleblower Heather Linebaugh, "It looks good on your resume if you kill more people."[642]

One person identifies the suspect by remotely accessing the suspect's front-facing camera. Another person launches and pilots the drone. Another person remotely aligns the drone's targeting system. Another person pushes the button to fire the missile. Another person confirms the identity of the dead body. By design, no one involved in the process feels responsible for a murder.[643]

On one occasion using a drone-launched air-to-surface missile, the U.S. government obliterated three U.S. citizens in Yemen, including a sixteen-year-old boy. No due process was provided. Of the three victims, only one was considered a terrorist. The other two deaths were labelled collateral damage.[644] "Collateral damage" is a term used to spin murder and manslaughter into something more palatable. It allows JSOC employees to continue seeing themselves as the good guys and rationalize remote-control murder.[645]

Collateral damage is a reoccurring theme with U.S. counterterrorism efforts. After Bin Laden was assassinated in 2011, it was publicly revealed that humanitarian workers were used as cover in an effort to find Bin Laden. The CIA embedded operatives in a nonprofit group called Save the Children. Subsequently, dozens of innocent humanitarian workers in Pakistan, suspected of being CIA agents, were collaterally murdered.[646]

2017: The Black Lives Matter Movement

The Black Lives Matter (BLM) movement drives home the point that a majority of black people feel as if their lives are undervalued. This is a by-product of power privilege. Power has been centralized in whites for centuries. Non-whites are all too aware of this. An overwhelming majority of U.S. politicians have been white. An overwhelming majority of U.S. police have been white. An overwhelming majority of U.S. businesses, including media, have been owned by whites and seem to prefer hiring whites. Stated differently, whites have had the privilege of being politically represented, policed, and hired by whites.

Because of movements like BLM, more white people are becoming conscious of the privileges built into this four-power structure. The U.S. Declaration of Independence states that "all men are created equal." It's a deductive expression of democracy. If white Americans support the Declaration of Independence and the U.S. Constitution, they will work to change the U.S. power structure into something more democracy and equality based.

Understanding the Right to Communicate: Kaepernick

By owning media corporations, white males have historically had power privileges involving media narratives. For example, when football player Colin Kaepernick kneeled for the national anthem during a National Football League (NFL) game in 2016, the point was lost on many whites.

After the game, Kaepernick clarified his intent: "I am not going to stand up to show pride in a flag for a country that oppresses black people and people of color. To me, this is bigger than football, and it would be selfish on my part to look the other way."[647]

The mass media narrative was then shifted by white male-controlled media to "respecting our troops," implying that kneeling during the national anthem disrespects U.S. soldiers who sacrifice their lives for their country. This misdirection was a distraction that prevented further communications involving accountability.

While it cost him his NFL career, Kaepernick's actions were more American than the flag itself. Nonviolently kneeling in front of the American flag on national television during the singing of the "Star-Spangled Banner" and pregame prayers was the perfect amalgam of U.S. First Amendment freedoms of religion, speech, assembly, and the press. It was the perfect opportunity to nonviolently communicate accountability information.

In later condemning Kaepernick, President Donald Trump tweeted, "Courageous Patriots have fought and died for our great American Flag—we MUST honor and respect it! MAKE AMERICA GREAT AGAIN!"[648]

To be clear, American patriots did not die for a flag. They died for the ideals America stands for, such as the freedom associated with power equality, including everyone making decisions for themselves rather than one person making decisions for everyone. The alternative is violence.

2018: A Microstudy in Centralized Power, the NFL

Power inequality in the NFL involves wealth, race, and gender. There are thirty-two teams. The owners are the decision-makers for those teams, except for the Green Bay Packers, which is owned by shareholders and governed by a president elected by the shareholders.

As of 2018, 100% of NFL owners were wealthy and white. Over 90% of NFL owners were male. This was the inevitable by-product of power privilege whereby white males had centralized power for centuries and therefore had the privilege of founding or purchasing NFL teams.

In a true democracy, if a town were 70% non-white, then its taxpayer-funded entities like police and judges would be 70% non-white. If a town were 50% female, then its police and judges would be 50% female. Representation would be *proportional*. Private businesses, however, are not democratic. They cannot be expected to operate like taxpayer-funded entities. No business owner wants someone else deciding who they hire. It would discourage entrepreneurship.

Consequently, the principle of proportionality does not apply to private businesses. If it did, then 50% of NFL players would be female simply because 50% of the U.S. population is female. If consumers do not like this fact, they can take their business elsewhere and create democratically run businesses. If the NFL truly supported equality, it would democratize the NFL's ownership structure, similar to the Green Bay Packers model. NFL fans could purchase shares.

Teams would be sold off to shareholders, with the sale proceeds going to the current owners.

An alternative solution would be for the NFL Players Association to start an entirely new football league with a power structure democratically controlled by the players. After all, the players are the key resource, not the owners.

2019: The MeToo Movement

The MeToo movement is a whistleblowing movement whereby people publish accountability information involving sexual misconduct. By working together, victims increase their collective communication power and are less likely to be overpowered by the resources of perpetrators. The movement has helped to equalize the power inequality between males and females as well as to illuminate the prevalence of sexual misconduct.

Sexual misconduct cases provide fertile soil for understanding power inequality. Almost everyone can relate to them. Hugh Hefner founded *Playboy* magazine in 1953 at the age of twenty-seven.[649] The magazine published photos of nude women and had a sleek bunny rabbit logo. Unlike publishers of human anatomy textbooks, Hefner was sexually objectifying women and was not protected by any democratic right to communicate photos accounting for the human body. Nevertheless, Hefner propagated the idea that he was a champion of free speech. It worked, and he became iconic: as he aged, he never strayed from having twenty-something-year-old girlfriends. The social mechanisms that made it possible have since been revealed by whistleblowers.

Hefner's business was initially quid pro quo. Women were compensated for nude photoshoots, simply trading visual sexual resources for financial resources. When the Playboy empire expanded to nightclubs in the 1960s, it featured waitresses known as bunnies. By rule, bunnies were under the age of twenty-five and wore uniforms essentially consisting of one-piece bathing suits, high heels, bowties, cotton-tails, and bunny ears.

Contractual Social Mechanisms

Being a bunny was lucrative at a time when the average female worker earned only 60% of the average wage for males.[650] Bunnies earned roughly twice as much money as waitresses working anywhere else.[651] Although it was an exploitation of power inequalities, the bunny employment arrangement was seemingly lawful. U.S. laws did not require equal pay for equal work.

At some point, however, an abuse of power was mechanized into the bunny employment agreement. Under the agreement, bunnies were given two options in the event they were, for example, sexually assaulted: (1) report the crime to Playboy security, or (2) not communicate the crime to anyone. Legally speaking, the clause was an obstruction of justice and an abuse of power by Playboy.

If a bunny were raped, the employment agreement required the victim to report the incident internally, centralizing accountability power in Playboy. Hefner then had the option power to blackmail the perpetrator. If the bunny decided to report the incident to police, her employment would terminate. Once reported internally, Playboy's "cleanup crew" would arrive at the scene of a crime and centralize accountability power by destroying evidence.[652]

Nightclub employment created a funnel for finding "playmates"—the girls photographed for the magazine. Seizing the opportunity, Hefner opened the Playboy Mansion, a Los Angeles residence where Heffner could host private parties with bunnies, celebrities, and other friends. The mansion was a magnet for both sexual and financial resources.

Mansion parties provided an implied tryout for girls wanting to become bunnies and for bunnies wanting to become playmates. At the same time, wealthy males could contractually become "keyholders," giving them access to both the nightclubs and the mansion. Keyholders were above the law while bunnies were below it. Hefner received daily reports accounting for any crimes involving bunnies or keyholders.[653]

Exposing the Playboy Empire

According to Joe Piastro, Playboy's head of security, one incident took place at the home of keyholder and television star Don Cornelius, the host of *Soul Train*. The bunnies involved were sisters, both roughly twenty years old. After being drugged and tied up, the two girls were repeatedly raped. After several days, one of the girls was able to free herself and telephone her Playboy supervisor, called a "bunny mother," who then directed the cleanup crew to pick up the girls at Cornelius's home. Consistent with their employment agreements, no one ever contacted the police. Cornelius was back at a Playboy nightclub the following week.[654]

Women at mansion parties were systematically drugged. According to Playboy employees, Hefner referred to quaalude pills as "thigh openers" or "leg spreaders."[655] According to former bunny P.J. Masten, married male guests would typically sodomize the bunnies to avoid getting them pregnant. Hefner had an elaborate hidden camera system at the mansion. It afforded him accountability power over guests. He could blackmail powerful people, including celebrities like singer Tony Curtis, movie producer Roman Polanski, and football star Jim Brown.[656] Actor Bill Cosby allegedly was caught on video molesting a fifteen-year-old guest.[657]

At the time, Los Angeles Police Department (LAPD) officials appeared to be abusing their power as law enforcement officers. Hefner began contracting retired LAPD personnel to work for Playboy security.[658] Active LAPD officers received perks in exchange for their cooperation with Playboy security. When a bunny reported a rape or some other crime to LAPD, the complaint presumably disappeared, and the bunny was fired. If the complaint reached some other law enforcement agency, the retired LAPD officers who worked as Playboy security could presumably use their law enforcement connections to make the investigation go away.

Sorority Syndrome

Hefner consciously created sorority syndrome to have sexual relations. Girls were deliberately turned against each other to compete for Hefner's affection. Heffner had the power to decide which girls had the options of: (1) magazine photoshoots, (2) rent-free residence at the mansion, and (3) cash allowances of roughly $1,000 per week.[659] No written contracts were technically needed. Some girls were allegedly raped, and some were not, but the machine worked like clockwork for Hefner's purposes.

By the time he was in his eighties, Hefner scheduled weekly orgies. He was the only male involved. Hefner gathered live-in and visiting girls in his master bedroom, where pornographic movies played in the background for stimulation. To maintain their place in the mansion, the veteran girls would pressure new girls to go over to Hefner. Resembling some form of psychological gang rape, the unsuspecting new girl was socially pressured into having unprotected sex with Hefner.[660]

Hefner had manufactured social mechanisms that provided him and his friends a steady stream of young girls, many of whom thought they were cleverly competing to be on the cover of *Playboy* magazine. In reality, they were at great risk of sexual assault by men undeterred by law enforcement. The dynamic was an extreme venue for power exchanges. The men had extreme financial and political resources, whereas the women had extreme sexual resources. Through exchanges of power with the LAPD, Heffner had minimized his risk. Through exchanges of power with girls, Hefner had maximized his reward. For Hefner's purposes, the governance system was optimal.

2020: Secret Ballot Elections

The White House is like the Playboy mansion of politics: Voters get sodomized, and campaign donors always win. Through exchanges of financial power with campaign donors, candidates purchase mass media communication power and maximize their connectivity with voters. For the 1%, this governance system is nearly optimal. U.S. elections come down to the same two options every time—Democrat or Republican. Campaign donors simply donate to both parties.

In the wake of both the 2016 and 2020 U.S. elections, the U.S. system of governance substantially destabilized. One side stood by the published election results, as is always the case, because its candidate won. The other side insisted the election was rigged, hacked, or stolen.[661] Both sides lacked accountability over voting results and could not technically make an informed decision as to who won.

The resulting instability culminated in a violent insurrection on January 6, 2021, when supporters of President Trump attempted a government overthrow.[662] This will inevitably happen again from utilizing a secret ballot election system in which vote-counting power is centralized in less than 1% of the population.

Until the 1880s, the U.S. utilized a decentralized election process. Each voter would approach a table where a clerk and sheriff were seated. In a transparent process, each voter would announce his name and the candidate he was voting for. The clerk would then document all the votes. Every voter could witness every vote.[663, 664]

Transparent elections ensured accountability over the election process. Other voters could tally the votes at the same time as the clerk. If the clerk's final voting results did not add up, voters could hold the clerk accountable. The decentralized accountability power provided by transparent elections helped prevent election rigging. Laws were in place to prevent retaliation from candidates, employers, or anyone else. For example, some colonies made it a crime for anyone to "menace, despitefully use, or abuse, any person, because he hath not voted as he, or they, would have had him."[665, 666]

By 1884, the election process began to change. On a state-by-state basis, Democrat and Republican Party politicians began shifting the election process from a transparent ballot system to a secret ballot system. Under the secret ballot system, votes were made in writing and dropped into a ballot box. The box then disappeared, and votes were counted by less than 1% of the voters.[667] The U.S. has remained a two-party political system ever since. The system is optimal for covert election rigging: two options and no decision transparency. By controlling options, political parties control decisions.

In the nineteenth century, many countries besides the U.S. made a similar shift to secret ballot elections.[668] Global power was being undemocratically centralized. It allowed the very people who were seeking reelection to have more control over vote counting. Election rigging became easier.

The political propaganda supporting the change touted "ending intimidation" at polling places. But rather than pass stronger laws prohibiting intimidation at polling places, party politicians instead centralized accountability power through secrecy.[669] Once voters become privy to election rigging, the default option is typically violence.

Democracy or Violence, Choose One

Political parties and campaign donations represent obvious conflicts of interest for political representatives. These representatives sit at the apex of power in global government and abuse the four powers by:

- **C:** leveraging their connectivity to solicit campaign donations
- **O:** optioning tax dollars for legislation that favors campaign donors
- **D:** voting along political party lines and providing favors to campaign donors
- **A:** protecting other politicians in their political party

Through cases like *Bank of the United States v. Deveaux*, *Buckley v. Valeo*, *Citizens United v. Federal Election Commission*, and *McCutcheon v. Federal Election Commission*, a Nixonized U.S. Supreme Court decided that corporations are people, money is speech, and campaign donations cannot be capped.[670, 671, 672, 673]

CONCLUSION

Centralized Power Is Heuristic

Centralized power is effective for small-scale governance and short-term problem solving. It works for smaller groups, such as parents governing families, teachers governing classrooms, or entrepreneurs governing small businesses. For purposes of large-scale and long-term governance, centralized power structures have failed. The risks have proven too great, as epitomized by the idea of one person having their finger on the "nuclear button." Centralized power is inevitably abused. As the saying goes, absolute power corrupts absolutely.

Abuse of centralized power is human nature, regardless of whether those empowered are white or non-white, Christian or non-Christian, male or non-male. Rather than repeatedly seeking out preferred group members and centralizing power in those group members, we have the option of creating decentralized autonomous organizations (DAOs) using technodemocratic social mechanisms.

Technodemocratic social mechanisms would allow for greater power balance, depending on the scale of governance needed across various groups and subgroups. In the case of large-scale and long-term governing bodies, DAOs would enable greater democracy and equality, as well as reduced social risk.

Technodemocracy: Decentralizing the Four Powers of Government

If you truly want to change the world, don't spin your wheels on specific problems such as global warming or nuclear proliferation. Start with understanding the system of governance. If you don't, a centralized system of governance will inevitably get in your way. Our modern systems of governance are rooted in centralized power and have been for millennia. The 1% does not want "change."

Technodemocracy is the principle of using technology to decentralize government power. It balances power to eliminate inequalities. Computer scientists, software engineers, developers, and programmers are now creating technodemocratic social mechanisms. At a high level, these mechanisms would enable the following:

C: *Decentralized peer-to-peer networks.* Each group member would be a network user directly connected to every other user through a mobile device, laptop, or some other computer network. The traditional Internet no longer would be necessary. Each network user's device could serve as a router, creating a network separate from the

Internet. Telephone calls, emails, texts, and the like would no longer pass through corporate middlemen such as AT&T, Verizon, or other network service provider corporations. Cell towers would no longer be needed. A technodemocratic network would be owned and governed by 100% of the users. In this system, political candidates would no longer have to purchase expensive ads from the communications middlemen of social media, television, or newspaper corporations.

O: *Decentralized peer-to-peer user identity verification.* Each user would have equal power to become a political candidate or to propose new options (ideas, products, processes, rules, laws, and so on). Candidates and voters would be connected directly to each other (no political parties). Information involving voting, currency, and other resources would move peer to peer and the risk of hacking would approach zero (the risk is inversely proportional to the number of active users). Resource access would be technodemocratic (equality based, not wealth based). For purposes of commerce, sellers would be connected directly with buyers using a shopping or marketplace blockchain (no Amazon or other middlemen). Transactions would occur using financial blockchain (no banks or other middlemen).

D: *Decentralized open-source decision-making software.* Decision-making applications such as voting apps would enable one person, one vote in regard to any candidate, proposal, product rating, and so on. Every major decision could effectively and efficiently be voted on. Like an Akashic record, blockchain-based software would provide eternal evidence of every election, resource transaction, or other major decision.

A: *Decentralized evidence-tracking software.* Traditional middlemen would be passively displaced using decentralized accounting software such as blockchain. Each user would be able to audit and account for every user, product, political candidate, voter, and vote in real time. Every user would also be able to audit and account for every timestamp and every bit of code for the underlying software in real time. Corporations would no longer govern and control marketplaces. Banks would no longer govern and control currency. Stated differently, accountability power would be evidence based (vested in 100% of users), rather than trust based (vested in 1% of users). Elected representatives would be accountable to voting blockchain (documented evidence of majority interests) rather than to campaign donors. In this way, technodemocratic representatives would gradually displace political party representatives.

For more information on theories of decentralized governance, see *Architecture of a Technodemocracy* (2018).

Blockchain-Based National Security

Like cryptocurrency, technodemocratic social mechanisms can now be built without the help of politicians, corporations, banks, churches, or any other middlemen. The possibility exists for blockchain-based national and global security. Much like the Internet has decentralized communication power, blockchain has decentralized accountability power. Decentralized cryptocurrency, cryptomedia, cryptomarkets, and cryptovoting have all become possible. One software developer can change the world.

A great portion of the accountability information in this book was not widely known until the twenty-first century. The Internet and mobile devices were not available prior to that time. These technologies have empowered people all over the world to connect and rapidly communicate information. We are entering an age of accountability.

The Building Blocks of Democracy: Social Mechanisms

Democracy is not something you ask for; it is something you build. The building blocks of democracy are *social mechanisms.* For millennia government social mechanisms have been autocratic, designed to maintain centralized power. After a handful of group members seize power, they inevitably abuse it by creating additional social mechanisms that further centralize power. For example, globally, tax laws generally burden the 99%, while members of the 1% use tax "credits" and in some cases pay no taxes at all.

Democratic social mechanisms help protect against power inequalities and abuses of power. Traditional examples include:

C: printing press (decentralized information)
O: marketplaces (decentralized resources)
D: elections (decentralized politics)
A: science (decentralized evidence)

Modern examples include:

C: mesh networks
O: mobile apps
D: mobile devices
A: blockchain

Cryptocurrency, for example, is designed to decentralize financial resources (banking). It is built on blockchain security software. Blockchain is like an Excel spreadsheet on steroids. Using transparent source code, decentralized blockchain software automatically calculates and transmits transaction decisions to 100% of its users, in real time, rather than centralizing that power in less than 1% of users, such as bank officials or corporate accountants.

Blockchain can be used to account for currency, social media, products, votes, people, contracts, legislation, and any other information resource.

Primary and Secondary Social Mechanisms

Democracy is a social or group phenomenon. It requires at least two group members. Otherwise, there is no one to overpower. A government is the primary social mechanism for constituting the group. The goal is group security through perpetual unification: a cyclical process of communicating, optioning, deciding, and accounting (CODA). An example of a primary social mechanism is the U.S. Constitution.

Group members are then accountable to the secondary social mechanisms documented in their constitution. Like software, secondary social mechanisms can be executed indefinitely. Democratic constitutions have historically decentralized power through perpetual social mechanisms such as elections (crowd voting), taxes (crowd funding), libraries (crowd educating), congresses (crowd legislating), military (crowd attacking), police (crowd investigating), and juries (crowd adjudicating).

History suggests that power tends to centralize naturally. Hoarding resources (centralizing option power) and imposing one's will (centralizing decision power) are timeless human attributes. Group members must endeavor to legislate secondary social mechanisms that counter and balance out power inequalities between group members (group members cannot violate a law that does not exist). An example of a secondary social mechanism is the U.S. Bill of Rights.

Governing power on Earth is currently centralized in less than 1% of group members—the middlemen. These middlemen spy on group members (centralizing communication power in themselves), hoard resources (centralizing option power in themselves), bribe (or "donate" to) politicians (centralizing decision power in themselves), silence whistleblowers (centralizing accountability power in themselves), or otherwise empower themselves to continue engaging in bad behavior.

Middlemen

Middlemen are the face of centralized power. They are the 1%. They are the politicians making critical social decisions without your input. They are the eavesdroppers surveying your most intimate communications for "threats." They are the corporations hoarding the resources you need to survive. They are the bankers burdening you with fees on resource transactions. They are the authority figures offering you spiritual or career advancement for sexual favors. They are the commercial and political propagandists distracting you from real accountability information. These middlemen stand between you and your fellow group members—your community. The 100%.

The middlemen, however, do not connect you to other group members. Technology connects you to other group members. Technology empowers you. The middlemen are disposable. Modern technologies can make full decentralization of all four powers possible:

C: decentralized networks, such as mesh networks, enable real-time connectivity
O: decentralized software, such as mobile apps, essentially provide lists of global resources
D: decentralized devices, such as mobile devices, offer hand-held decision-making
A: decentralized ledgers, such as blockchain, account for the decisions of all group members

These four technologies are the pillars of a technodemocracy. They create a platform that enables us to bypass the middlemen. A technodemocratic platform would be 100% user funded, owned, and controlled. There would be no corporate, bot, or otherwise nonhuman user accounts. One person, one account. No fees. No advertisements. No surveillance.

A transition can now be made from, for example, central banks, political parties, and corporate social media to decentralized peer-to-peer social mechanisms, such as cryptocurrency, cryptomedia, cryptomarkets, and cryptovoting. The peer-to-middleman-to-peer model is no longer our best option.

While the four powers of government in most countries have been slightly decentralized from individual group members, such as kings and dictators, they still remain centralized in less than 1% of the population, such as corporations and politicians. A revolution that decentralizes the four powers from the 1% to the 100% is underway. Computer scientists, software engineers, developers, coders, programmers, and technodemocratic leaders are at the cutting edge of that revolution. The next generation of revolutionaries will be armed with computers.

Mobile Devices that Function as Mini Cell Towers

The Internet is predominantly undemocratic for the same reason modern republics are predominantly undemocratic. Power is vested in a handful of users (the 1%) instead of being vested in all users (the 100%). Yes, the Internet is democratic in that it connects people all around the world. Communication power has been drastically decentralized by the Internet. However, options, decisions, and accountability over the Internet are still centralized in corporations like Verizon, AT&T, and Century Link, as well as in the politicians funded by those corporations.

A fully decentralized communications network requires mechanisms offering peer-to-peer networking services. Mobile devices that displace Internet service providers and function as mini cell towers must be invented. By protocoling data to hop from user to user, instead of cell tower to cell tower, the corporate middlemen can be cut out. Verizon, AT&T, Century Link, and the like would no longer be needed.

Decentralized Blockchain vs. Centralized Blockchain

Corporations like International Business Machines (IBM) are creating *centralized* blockchains. Centralized blockchains allow the hosting group member or corporation to have centralized accountability power. Centralized blockchains are intended to preserve the middlemen. The middlemen can charge a fee, impose advertisements, or conduct warrantless surveillance of users. These centralized blockchains are an undemocratic attempt to maintain centralized power by hijacking modern blockchain technology.

Artificial Intelligence Provides Options, Humans Provide Decisions

As discussed in Part IV of this book, accountability power is typically centralized in government agencies and corporations by way of self-regulation. For example, the U.S. Federal Aviation Administration (FAA) historically has allowed airline corporations to self-regulate aircraft safety measures. This is undemocratic and results in unpoliced conflicts of interest. Stated differently, aircraft manufacturers can cut corners to save money and increase profit.

Consider governance over an aircraft:

- **C:** the aircraft's parts are all connected and potentially in communication with the pilot
- **O:** each part offers resources, such as fueling, propelling, accelerating, stabilizing, and braking

D: the pilot (whether human or technological) makes decisions

A: accountability occurs through pilot sensation, speedometers, altimeters, compass, and the like

In a technodemocratic system, artificial intelligence (AI) software is utilized to create more options, not to centralize decision power. If AI is used to make decisions, decision power becomes centralized in the corporations and software developers creating that technology. For example, in an aircraft, AI would be useful for providing a pilot with real time options, such as optimal course corrections in light of inclement weather. A human pilot can then make decisions.

If a computer's AI undemocratically makes the decisions, the plane might be programmed to always take the most fuel-efficient course. This programming would reduce fuel costs and increase profits for airline corporations. However, if that AI-proposed course fails to factor in exigent circumstances, such as a damaged sensor or hostile airspace, the outcome could be catastrophic.

Philosophically speaking, a human pilot, not computer software, should always be vested with decision power. This is why AI software developers, and engineers in general, need to fundamentally understand the four powers and how they apply to any particular mechanism, be it social, nonsocial, or anything in between. For purposes of a democracy, AI can be used to empower a user with more options, but the user should still make the decision.

Decentralized Voting Blockchain

All the resources in the universe derive from one source—the creator of the universe. Our creator is our only source of power. How we decide to utilize those resources is within our power. When we make decisions as individuals, we change the universe. When we make decisions as a group, we vote. As such, voting is at the apex of universal social change.

To provide mathematical perspective, consider a monarchy. In a monarchy, if a king rules over a kingdom of 400 million people (including himself), the king has 400 million voting units. Outside the king, each group member has zero voting units. The king makes all the governing decisions. If the king is democratic, he will communicate with group members, tally their opinions, and make decisions consistent with majority interests. If the king is undemocratic, he will do whatever he wants. Either way, accountability power is undemocratically centralized in the king. Simply put, the king is the only one accounting for the votes.

Alternatively, consider a republic. If the U.S. has a voting population of 400 million people, then theoretically each person has one voting unit. No person has 400 million voting units. Under the U.S. Constitution, these voting units only vest during elections every two, four, or six years. The rest of the time, the American people are disempowered. Because a republic is a representation-based system of government, representatives are empowered the rest of the time.

Elected U.S. representatives have 435 votes in the House and 100 votes in the Senate. That works out to about 919,000 voting units (400 million divided by 435) for each House representative and 4 million voting units (400 million divided by 100) for each senator. These representatives undemocratically do not communicate with their constituents once elected. On important legislative decisions, there is no evidence of majority interests. Subsequently, U.S. representatives are unaccountable for decisions made on their respective congressional floor.

If majority interests were documented, we would have the evidence needed to hold elected representatives perpetually accountable. This is gradually becoming possible using decentralized voting technologies. Decentralized voting technologies are the cornerstone of technodemocratic governance:

C: voting only requires bytes of data, zeros ("no" votes) and ones ("yes" votes)
O: users see every proposal or option in real time
D: users see every vote or decision in real time
A: voting blockchain provides *documented evidence of majority interests* (DEMI)

In a technodemocratic republic, DEMI would be used by elected representatives when voting on their respective congressional floor, as well as by voters seeking to hold their representatives accountable. Elected representatives would be accountable to DEMI rather than campaign donors. In a technodemocratic system, political parties, campaign donors, and corporate lobbyists would gradually lose their influence. They would be phased out, due to the arrival of technodemocratic representatives.

For more information on decentralized voting technologies, see Democracy Earth Foundation, Horizon State, Follow My Vote, Voatz, Agora, Polys, Luxoft, Votem, Polyas, B-Vote, Boule, and Flux.

Decentralized Smart Contract Blockchain

Creating contracts is a democratic process. When you make a formal decision as a group of two or more, it is generally a contract. All parties to the contract have a vote in deciding to form the contract. By using smart contract blockchain, agreements become more dynamic. To better understand the benefits of blockchain, consider the requirements for a smart contract involving a beverage vending machine:

C: the vendor's machine communicates the options (drinks, prices, payment methods)
O: the vendor's machine actively supplies drink options (water, soda, coffee)
D: the purchaser decides on a drink and payment, and the vendor's machine decides to dispense
A: the transaction is documented on a smart contract blockchain

If the vendor's machine does not dispense as agreed, the purchaser effectively must blow a whistle on the vendor. Unlike with traditional vending machines, wherein the purchaser might lose the payment and has no evidence of actually paying, the smart contract provides documented evidence that the purchaser fulfilled his or her end of the bargain. This is a subtle distinction that decentralized blockchain security provides (decentralized evidence). The group is no longer required to trust a middleman. In the case of a smart vending machine, payment can be autonomously and remotely refunded by the community without the vendor's consent.

Consider the contract of marriage. In the case of smart marriage contracts, marriage requires two votes, and divorce requires one. Couples could easily be required to document a basic divorce agreement as a condition of getting married. This new social mechanism would serve the public interest by preemptively reducing divorce litigation. Communities would not have to trust middlemen (such as feuding parents) to ensure the security of children in that community. Determining parental custody, asset distribution, and so forth would be a condition of marriage.

Decentralized Legislative Process Blockchain

In a technodemocratic system, all the major resources at the disposal of the group would be accounted for by blockchain, including currency, social media, products, votes, people, land, food, water, precious metals, medicine, and so on. Individual group members would no longer be able to manipulate life-essential resources through hoarding, capitalizing, or destroying. This chain-of-custody accountability allows greater sustainability, efficient processing, and more effective use of resources by decision-makers. Both data analytics and AI could be leveraged to anticipate shortages, economic harm, environmental harm, or other social problems. Social risk could be drastically reduced.

Consider the manipulation of healthcare resources. During the coronavirus pandemic of 2020, some U.S. politicians were avoiding business closures because it would slow the economy and presumably harm their chances of reelection.[674] In a technodemocratic system, legislation involving business closures would be decided democratically, based on health risks and economic risks rather than political risks. Some campaign donors, particularly pharmaceutical corporations, wanted vaccines mandated.[675] In a technodemocratic system, legislation involving vaccines would occur democratically, based on scientific evidence and documented evidence of majority interests (DEMI) rather than on corporate profitability.

Decentralized Executive Process Blockchain

In theory, the upside of perpetual surveillance is reduced crime. In reality, group members controlling such surveillance have the centralized accountability power to cover up their own crimes (i.e., the CIA). This is the downside of having cameras on every street corner in a system where governing power is centralized. In a technodemocracy, cameras on every street corner, ambulance, patrol officer, patrol vehicle, and patrol weapon could potentially be streamed to a municipal blockchain.

In the case of prison systems, for example, each inmate's location of incarceration, time in, time out, and so forth would be available on a municipal blockchain. No one disappears. A municipal blockchain would allow for the democratic monitoring of life-essential resources, including people, food, water, energy utilities, and medical supplies, as well as provide decentralized accountability over publicly funded cameras and other evidence-recording mechanisms.

Through proper computer coding, a municipal blockchain could be remotely unlocked by:

(1) local and federal prosecutors and politicians, in real time, or
(2) majority interest from the citizens within that jurisdiction.

Until municipal recording technologies are controlled by 100% of the group, these accountability mechanisms are susceptible to abuses of power.

Decentralized Judicial Process Blockchain

In a technodemocratic system, judicial mechanisms such as judges and juries are designed to reflect the corresponding community. They follow the democratic principle of proportionality. A technodemocracy enables power equality through social mechanisms that dynamically adjust municipal bodies to remain perpetually equal to a cross-section of their respective communities.

For example, if the community is 40% white, 40% black, and 20% other, then judges and juries would be approximately 40% white, 40% black, and 20% other. If the community is 48% female and 52% male, then judges and juries would correspond to those proportions.

Decentralized Identity Blockchain

Identity blockchain would account for nationality, citizenship, and voter registration, as well as marriages, business partnerships, and every other alliance. In a broader sense, it would function as an alternative system for partnership accounting without the need for middlemen, such as dating apps, job apps, networking apps, shopping apps, other corporation-controlled social media, marriage records, government birth certificates, social security administrations, passport servicers, corporation commissions, or other government administrators.

An identity blockchain would offer a user-controlled, public-facing profile similar to Facebook, Amazon, or LinkedIn, but without the corporate middleman. In the case of dating or sales, Person X empowers Person Y by giving Person Y the option to passively connect with Person X at a later date, on Person Y's terms instead of Person X disempowering Person Y by interfering with what Person Y is doing at the moment. Nobody gets hit on. Nobody gets force-fed ads or any other propaganda.

Decentralized Financial Blockchain

Traditionally, banks have accounted for their transactions using a centralized bank ledger such as an Excel spreadsheet. A central group member, such as an accountant or bank official, serves as a middleman who has the accountability power to fraudulently "fudge the numbers." Depositors trust this middleman with their life savings.

In 1776 Scottish economist Adam Smith published the book *Wealth of Nations*, a capitalist manifesto that championed an every-man-for-himself system. In that book, Smith theorized that an "invisible hand" serendipitously guides society to economic prosperity when individuals pursue personal interests over majority interests. Smith used the Dutch Bank of Amsterdam as a shining example of the invisible hand at work.

In the 1770s the Bank of Amsterdam became more popular than other banks because it did not make loans. It merely charged its customers transaction fees, such as when they opened accounts or made withdrawals. The Bank of Amsterdam was very profitable while providing a low-risk, deposit-only service to the people of Amsterdam. Smith described it this way: "The bank of Amsterdam professes to lend out no part of what is deposited with it, but for every [deposit] for which it gives credit in its book, to keep in its repositories the value of a [deposit] either in money or bullion. That it keeps in its repositories all the money or bullion for which there are receipts in force, for which it is at all times liable to be called upon…"[676]

One day, in 1790, when depositors went to the bank to make withdrawals, their money was no longer there. Bank officials were secretly making loans and pocketing the interest and fees. When borrowers stopped making payments on the loans, the Bank of Amsterdam declared itself insolvent, and depositors had no recourse. The "invisible hand" of the 1% had stolen their trust.[677]

A platform is not democratic unless it is perpetually governed by 100% of its users. In a technodemocratic system, blockchain-based platforms automatically communicate a copy of the ledger to every user after every transaction in real time. In this way, every user is connected. Users can *communicate* and *account* for every *option* and every *decision*. There is no corporate middleman (i.e., capitalism). There is no government middleman (i.e., socialism or communism). All four powers are decentralized to 100% of the users.

The Most Valuable Cryptocurrency on Earth

By virtue of being the first cryptocurrency to market, Bitcoin's developers empowered themselves with centralized option power (cryptocurrency resources). Simply put, they compensated themselves with Bitcoin. The primary inventor, who is yet to be identified (known only by the pseudonym Satoshi Nakamoto), may eventually become the world's first trillionaire. This is the problem with building any blockchains on top of the Bitcoin platform. It is not a decentralized cryptocurrency.

To a lesser extent, the developers of the Ethereum cryptocurrency also centrally empowered themselves. Decision power involving changes to Ethereum's underlying software is centralized in a handful of founding users and software developers rather than in 100% of Ethereum's users. Like Bitcoin, Ethereum's builders also have centralized option power, owning large quantities of Ethereum.

Theoretically, the most valuable cryptocurrency on Earth will be the one that successfully decentralizes all four powers of government. Like owning stock in a decentralized version of Facebook, Amazon, the United Nations, and WikiLeaks all rolled into one, it will be far more valuable than Bitcoin or any other cryptocurrency to date. It will be backed by 100% of its users. It will be socially optimized.

APPENDIX A

Advice for Whistleblowers

The four powers framework ultimately illuminates the primary rights required within any democratic system of governance:

C: the right to communicate (religion, speech, press, assembly, protest, etc.)
O: the right to options (food, water, shelter, healthcare, energy, information, etc.)
D: the right to decide (living, dying, marrying, reproducing, immigrating, voting, etc.)
A: the right to accountability (whistleblowing, auditing, investigating, impeaching, etc.)

One way to express the right to accountability is to whistle blow. Generally speaking, whistleblowing is dangerous. Like a referee at a sporting event, someone will hate you for blowing the whistle. Seizing your right to accountability comes at a price. You are empowering yourself, but you are disempowering someone else. You are disrupting the status quo and becoming a threat. You will likely be scrutinized, criticized, humiliated, demoted, fired, criminally prosecuted, imprisoned, physically harmed, or even killed. That's the easy part.

The hard part is the ripple effect that follows—feelings of powerlessness, depression, and anger; thoughts of suicide or violence; distrust of authority; destabilized relationships; sudden unemployment; loss of health insurance, career options, job references, credit, and security clearances; legal expenses; student loan default; foreclosure, bankruptcy, and court-liquidated retirement funds; and credit–based withdrawals of employment offers. The ramifications last for decades.

In August of 2004, I was hired by the DOJ as an FBI intelligence analyst. After two months I was promoted from pay grade GS-9 to GS-11. After three months, I was recruited as a counterterrorism agent because (as I was told by the new agent recruiter Manny Johnson) FBI Headquarters, following 9/11, was "hiring engineers to work counterterrorism." Before leaving for new agent training in May 2005, I was provided tens of thousands of dollars in taxpayer-funded training regarding intelligence collection, cyberattacks, counterintelligence, and counterterrorism.

In October 2005, I was assigned from new agent training in Quantico to the FBI Honolulu drug squad. The amount of dysfunction was breathtaking. On the morning of my first day, my "training agent" introduced me to the FBI Honolulu Special Agent in Charge (SAC) Charles Goodwin and an assistant SAC (ASAC), Robert J. Casey. SAC Goodwin was on his way to play golf, and ASAC Casey gave us advice about Honolulu real estate investing. The training agent then left to run errands.

During my entire time in Honolulu, I was never told the mission of the FBI Honolulu field office, the goals of the drug squad, or what my role would be. There was no sense of direction.

The training agent told me to put 8:15 a.m. to 6:15 p.m., Monday through Friday, on all my time and attendance cards or else I would not get paid. The training agent also told me to buy several Hawaiian shirts and show my face by 10:00 a.m. each morning.

The training agent said I could do whatever I wanted but had to make sure my phone was always on. If anyone were to ask, I was to tell them I was driving around learning the streets of Hawaii. And if I wanted to look like I was doing work, I was to turn in one sheet of paper each day.

I saw the training agent about once a week. Because the drug squad area was overcrowded, the training agent gave me a desk on the cyber squad. This effectively made me invisible to the drug squad. SA Arnold Laanui was on the cyber squad and sat at the desk across from mine. He affectionately referred to the drug squad as "the squad that doesn't do any work."

My work dynamic immediately became unnecessarily stressful. The cyber squad was polite but awkwardly would ignore me because I was not part of their team. I would walk over to the drug squad area, see that no one was there (except for the squad secretary), and then walk back to the cyber squad area. This went on for months. Consistent with FBI new agent training, I kept my mouth shut.

After the first month, I tracked down my immediate supervisor, Supervisory Special Agent (SSA) Dan Kelly. I told him my background was engineering and intelligence collection and that instinctively I was not a cop. I needed mentoring to be effective on a drug squad.

After requesting a mentor, SSA Kelly directed me to SA Tim O'Malley. When I asked SA O'Malley what I needed to do to become a top agent, he told me to let the senior agents come to me. In December 2005, a senior agent came to me. At the FBI Honolulu Christmas party, I was sexually harassed by SAC Goodwin on behalf of ASAC Pam McCullough, who apparently wanted to give me her hotel room key. I declined by ignoring them. I was immediately off on the wrong foot with the two most powerful FBI agents in the state of Hawaii. In the following weeks, SAC Goodwin and ASAC McCullough were both avoiding me, putting their head down and changing direction, or blushing with embarrassment, respectively.

The red flags kept on coming. After 9/11, FBI Headquarters had been shifting resources to counterterrorism and had prohibited field offices from opening new drug investigations. When I asked SSA Kelly how the Honolulu drug squad had been opening new drug investigations, he confessed that he was piggybacking new drug cases on existing drug cases by opening them as "sub-cases," which was "easy to do on an island." I also asked why the drug squad was assigned more agents than the counterterrorism squad. I was told FBI officials get cash bonuses for asset forfeiture, and drug cases result in more asset forfeiture than counterterrorism cases. This was a clear conflict of interest.

The communication in Honolulu was terrible. After two months, the training agent realized I was not on the drug squad email lists. I was out of the loop on, among other things, my first arrest. I never saw a warrant and was simply told by phone where to go. A couple teenagers were being arrested at a crowded shopping mall in a joint effort with the Honolulu Police Department. The planning appeared to be improvised and word-of-mouth. I was never introduced to any of the Honolulu police officers or other FBI agents who showed up. Everyone was in street clothes. I could not distinguish who was a cop and who was a civilian. I was simply told to point my gun at the pregnant teenage passenger in the suspect's vehicle. Consistent with FBI new agent training, I kept my mouth shut.

On another occasion, I was directed by a senior agent to interview an inmate at a local prison. The senior agent did not go with me. During the interview, the inmate appeared to be sizing up whether I had a gun under my Hawaiian shirt. A few days later, I was told by the training agent that FBI guidelines had been violated because I had conducted the interview without at least two agents present. The leadership failures were becoming physically stressful. I started to experience health problems, including stomach cramps and sleep loss. Consistent with FBI new agent training, I again kept my mouth shut.

FBI Honolulu was also financially stressful. Under new agent employment contracts, the FBI pays relocation expenses, such as airfare, temporary housing, a rental car, and shipping costs for personal property. The exceptions are relocations to Hawaii and Puerto Rico. When a new agent is assigned to Hawaii or Puerto Rico, the new agent pays relocation expenses, and the FBI contractually agrees to reimburse those costs.

I paid my Honolulu relocation expenses using a credit card. My reimbursement paperwork sat on my drug squad supervisor's desk for months (I could see it under a pile of paperwork) and was presumably never sent to FBI Headquarters for processing. To this day, I have not been reimbursed by the FBI for more than $10,000 in relocation expenses. (The credit card company was reimbursed two years later after a bankruptcy court liquidated my 401K retirement fund.)

During my first four months as an FBI agent, I was tasked by random senior agents to do, at most, eight hours of work per week. In February 2006, I became a whistleblower under 5 U.S.C. §2303 by communicating the following protected disclosures to the "highest-ranking FBI official" in the state of Hawaii (SAC Goodwin): (1) I had seen my supervisor SSA Kelly only four times in four months (evidencing time and attendance fraud), (2) the drug squad had zero meetings (evidencing gross mismanagement), and (3) I did not know what I was supposed to be doing (evidencing a lack of "vigilant oversight and direction," as required for probationary new agents under FBI MAOP §21-1).

SA Henry "Hank" Hanburger, an agent on the white-collar crimes squad, volunteered as an intermediary and arranged a private meeting between me and SAC Goodwin. SA Hanburger was a friend of SAC Goodwin.

SAC Goodwin, however, never showed up for that meeting. As an attorney and as the highest-ranking FBI official under 5 U.S.C. §2303, he was likely attempting to obstruct additional protected disclosures and short-circuit the statute all together. SAC Goodwin's secretary escorted SA Hanburger and me out of SAC Goodwin's office to the office of ASAC Casey. I ultimately proposed to ASAC Casey that I be reassigned to another squad. He refused. I then proposed I be transferred back to my previous position as an intelligence analyst in Phoenix. He appeared insulted that someone would rather be an intelligence analyst in Phoenix than an agent on the Honolulu drug squad. That being said, ASAC Casey expressly agreed, without objection, to transfer me back to Phoenix and to keep the meeting confidential.

Two days after I blew the whistle, drug squad agents started showing up for work. Despite the confidentiality agreement I had with ASAC Casey, drug squad agents had been informed that I was trying to get off their squad. Some of them were not happy I had drawn attention to the squad. A few began lecturing me, others gave me dirty looks or ignored me. ASAC Casey subsequently showed up in the cyber squad area and announced to everyone within earshot that Phoenix had received its FSLs (Funded Staffing Levels) and implied that I would soon be transferred back to one of four intelligence analyst openings in Phoenix.

Two weeks after my whistleblowing, my squad supervisor, SSA Kelly, was replaced by SSA Ed Arias. I was told by SSA Arias I was being transferred back to Phoenix.

Four weeks after whistleblowing, I walked into the building at around 7:30 a.m. as if it were just another day. Presumably acting at the direction of SAC Goodwin (who had conflicts of interest in view of both 5 U.S.C. §2303 and sexual harassment laws), my new supervisor, SSA Arias, escorted me into the office of ASAC Casey. I was surrounded by four armed senior agents. Everyone was presumably following SAC Goodwin's orders. I was then effectively ordered to sign my own resignation letter.

ASAC Casey threatened me with a criminal investigation for time and attendance fraud if I did not sign the so-called resignation letter he had prepared in my name. No notice nor due process was provided. My heart rate spiked. I felt overpowered. ASAC Casey was abusing his power to open criminal investigations as leverage to force my resignation. SA Hanburger began arguing with ASAC Casey. The situation deescalated when I surrendered my firearm and involuntarily produced a signature.

As I tried to process what was unfolding, the security escort let me say goodbye to a few people. While cleaning out my desk, it felt like the entire office paraded by to see what was going on. It was heartbreaking and humiliating. I was holding back tears. Because I suspected that I would be accused of stealing government property, I left all personal property (including FBI employment and relocation agreements) in boxes under my desk. I genuinely feared further retaliation. I later learned from a former FBI coworker that, after whistleblowing, I had been placed under surveillance by my superiors.

During my entire time at the FBI, all my employment evaluations had been positive. Being fired was a direct result of whistleblowing. I was never provided due process notice that I was doing anything wrong or at risk of being fired. The training agent had told me to put 8:15 a.m. to 6:15 p.m., Monday through Friday, on all my time and attendance cards, regardless of hours worked. The drug squad secretary was a witness.

By controlling my options, my superiors controlled my decisions. None of my options were good. I could: (1) refuse to sign a fraudulent resignation letter written by someone else and be unobjectively criminally investigated by my direct supervisors, (2) use my gun, or (3) simply sign, albeit under duress. ASAC Casey abused his power under color of law (by threat of a criminal investigation). Additional "resignation paperwork" was apparently forged and mailed to FBI Headquarters by SAC Goodwin.

I later learned from a former FBI coworker that I was rumored to have been fired for sleeping with Honolulu ASAC Pam McCullough. I was never actually introduced to ASAC McCullough and certainly did not sleep with her. At the 2005 FBI Honolulu Christmas party, held at the Hilton Waikiki, a possibly intoxicated SAC Goodwin had offered me her hotel room key while ASAC McCullough simply stood behind him and blushed, seemingly not expecting him to make such a remark. By no act of my own, I had become a threat to the two highest-ranking FBI officials in the state of Hawaii. I suddenly had the option power to file a sexual harassment lawsuit against the FBI, something I had no intention of doing.

Filing such a lawsuit would have created the public perception that SAC Goodwin was gay (a twenty-something-year-old male new agent would be filing a sexual harassment lawsuit against a married male FBI official). In addition, SAC Goodwin had been offered a high-paying job with the Professional Golfers' Association (PGA) Tour. Six months away from his scheduled retirement, he may not have wanted to risk losing his current job, his retirement pension, or his next job. He had conflicts of interest. To a reasonable observer, the decision to terminate would appear to be an abuse of power.

Firing me without due process, such as a standard OPR (an internal investigation conducted by the FBI's Office of Professional Responsibility), allowed my superiors to avoid having their own actions accounted for by FBI Headquarters. In addition, my only source of income was instantly cut off. I could no longer afford an attorney (typically at least $300 per hour with a $10,000 or more retainer). My credit was already maxed out from $10,000 in unreimbursed FBI Honolulu relocation expenses.

In 2009 after overcoming sudden loss of income, unemployment, emotional distress, fear of additional retaliation, liquidated savings, student loan default, foreclosure, homelessness, bankruptcy, court-ordered liquidation of my retirement fund, and the inability to afford legal representation, I filed a sixteen-page whistleblower retaliation complaint with the DOJ inspector general. The DOJ inspector general did not respond but instead tipped off the FBI. I was then contacted by FBI General Counsel Valerie Caproni. Caproni told me I would be rehired and wanted me to come in for an "employment polygraph."

In reality, the polygraph was an interrogation. The FBI polygrapher was condescending and admitted he had read my whistleblower retaliation complaint (the one confidentially filed with the DOJ inspector general). When I told the polygrapher that my squad in Honolulu had committed time and attendance fraud, he said, "I wouldn't have used those words," and reclined as if I had just incriminated myself. Finally, when I refused to disclose the names of FBI employees still in contact with me (presumably leaking information to me supporting my lawsuit), the polygrapher threatened that I would not be rehired if I did not disclose. I refused to disclose.

In response, I stopped communicating with any and all friends I had made at the FBI. Shortly after the polygraph, I received a letter from FBI Human Resources informing me I would not be rehired. Over the phone, I was called a liar by my FBI Human Resources point of contact. I then finally received a cookie-cutter first letter from the DOJ inspector general's office informing me there would be no whistleblower retaliation investigation. The letter cited "limited resources" as a justification for not enforcing the FBI whistleblower statute under 5 U.S.C. §2303. No witnesses were ever contacted.

When I petitioned my so-called representatives, including members of the U.S. Senate and House intelligence committees, nobody responded.

In 2010 I filed *Hanania v. Office of the President of the United States* in an Arizona federal court. The president is expressly responsible for enforcement of the FBI whistleblower statute under 5 U.S.C. §2303. Within that lawsuit, I requested the judicial branch order the executive branch to open a whistleblower retaliation investigation. Once a lawsuit was filed, DOJ attorneys suddenly began to communicate with me. They did not offer to investigate the facts. Instead, they threatened me with criminal prosecution for disclosing classified information and called me "disgruntled."

To their embarrassment and my luck, the so-called classified information in question had been cut and pasted from the FBI's own website. Nevertheless, in violation of separation of powers principles, an FBI attorney entered the federal courthouse and physically destroyed all copies of my lawsuit. The FBI had erased any public record or account of my experience. Anticipating harsher retaliation, I then emailed copies of the lawsuit to roughly thirty active FBI employees. I also refiled the lawsuit in a California federal court, although the lawsuit was eventually dismissed.

Accountability is an evidence collection process. In my case, the DOJ had the luxury of investigating its own crimes. The DOJ inspector general failed to telephone witnesses, review documents, collect other evidence, or otherwise "inspect" the FBI. Simply put, the DOJ inspector general's office had centralized accountability power and conflicts of interest. It was protecting the FBI brand, minimizing DOJ liability, and reducing its workload instead of fulfilling its duty to inspect under 5 U.S.C. §2303. The decision not to inspect was an abuse of power.

I would like to thank the thousands of whistleblowers worldwide who made this book possible. If you are a government employee contemplating whistleblowing, do not tip off your superiors. They have accountability power over evidence and witnesses. They also have the decision power to retaliate, ending your government career instantly and cutting off what may be your only source of income. The ramifications are lifelong.

As inspired by Edward Snowden, consider the three A's:

(1) **Anonymity:** Do not whistle blow to the accused entity. Focus on collecting evidence that proves your claims, preferably without it being traceable to you. Documents, audio recordings, photographs, and the like are worth a thousand words.
(2) **Attorney:** Covertly meet with an attorney. Use the attorney as a sounding board, and possibly have him or her release evidence on your behalf as their "anonymous client" (you are legally shielded by attorney-client privilege). The right attorney will likely represent you at no cost.
(3) **Amplification:** Ask your attorney to connect with other attorneys, whistleblowers, journalists, and civic leaders who have the communication power needed to rapidly disseminate accountability information to community members positioned to solve the problem you expose.

In addition, consider reading *Caught Between Conscience & Career*, which can be downloaded from the Project on Government Oversight website at www.pogo.org.

APPENDIX B

CODA Quick Reference Guide

COMMUNICATION POWER

- *News:* contemporaneous accountability information (e.g., speech, press, voting results, judicial decisions)
- *Education:* non-contemporaneous accountability information (e.g., books, documentaries, historiographs)
- *Propaganda:* deceptive information propagated to influence decision-making (e.g., spin, lies, false flags)
- *Other:* any and all other null information (e.g., noise, distractions, pornography)

OPTION POWER

- *Natural resources:* food, water, land, horsepower, oil, electricity, nuclear power, other energy
- *Human resources:* labor, sexuality, spirituality, creativity, persuasiveness, humor
- *Financial resources:* currency, loans, securities, taxes, other crowdfunding, precious metals
- *Information resources:* voting results, legislated rights, judicial decisions, publications

DECISION POWER

- *One ruler (king, dictator, employer)*: One voter (democratic only when the ruler empowers the people with CODA)

- *The 1% (representative-based republic)*: Decision power can be purchased (i.e., bribery is legalized through campaign donations)
- *The 100% (equality-based)*: Decentralized peer-to-peer decision-making (i.e., mesh networks, mobile apps, mobile devices, and blockchain)
- *Evidence*: Centralized (such as in politicians or corporations) versus decentralized (such as publicly documented blockchain)

ACCOUNTABILITY POWER

- *The 1% hold the 1% accountable*: Checks-and-balances process (U.S. Constitution, Articles I–III, separation of powers)
- *The 1% hold the 100% accountable*: Legislative, executive, and judicial processes (U.S. Constitution Articles I–III)
- *The 100% hold the 1% accountable*: Nonviolently or violently (U.S. Constitution First or Second Amendment, respectively)
- *The 100% hold the 100% accountable:* Using an evidentiary process, such as due process (U.S. Constitution Fifth Amendment)

APPENDIX C

CODA Applicative Example: the U.S. Legal System

COMMUNICATION POWER

- *First Amendment:* The right to communicate accountability information through spirituality, speech, press, assembly, and petition
- *Second Amendment:* The right to bear arms, including guns, networks, cryptocurrency, mobile apps, mobile devices, and blockchain
- *Government centralization:* Creating secrecy laws and covert surveillance agencies (CIA, NSA, FBI)
- *Corporate centralization:* ABC, NBC, CBS, Comcast, AT&T, Verizon, T-Mobile, Google, Facebook, the New York Times

OPTION POWER

- *Natural resources:* property laws, environmental laws, incorporation laws
- *Financial resources:* banking laws, currency laws, securities laws, tax laws
- *Human resources:* employment laws, incorporation laws, military draft laws
- *Information resources:* secrecy laws, data privacy laws, intellectual property laws

DECISION POWER

- *Article I:* legislative processes (assembly, floor voting, veto override, impeachment)
- *Article II:* executive processes (commander-in-chief of military, executive orders, vetoes)

- *Article III:* judicial processes (evidence discovery, juror selection, adjudication, sentencing)
- *Articles I–III:* election processes (candidacy requirements, representation terms, appointing electors)

ACCOUNTABILITY POWER

- *Articles I–III:* separation of legislative, executive, and judicial functions (implying accountability through checks and balances)
- *First–Tenth Amendments:* the right to accountability. See *Architecture of a Technodemocracy* (2018), Chapter 6
- *Government centralization:* politician-controlled (rather than citizen-controlled) elections, law enforcement, and courts
- *Corporate centralization:* self-regulation, incorporated ratings agencies, and patent and trade-secret protections

APPENDIX D

CODA Student Exercises: the U.S. Bill of Rights

After answering each question, consider why you came to your conclusion.

1. Communication power in the context of the First Amendment:

   "Congress shall make no law respecting an establishment of religion, or prohibiting the free exercise thereof; or abridging the freedom of speech, or of the press; or the right of the people peaceably to assemble, and to petition the government for a redress of grievances."

   - Are unspoken spiritual communications protected?
   - Are spoken communications protected?
   - Are printed or published communications protected?
   - Does assembly of the group, without additional action, constitute a protected communication?

2. Option power in the context of the Second Amendment:

   "A well regulated militia, being necessary to the security of a free state, the right of the people to keep and bear arms, shall not be infringed."

   - Must First Amendment rights be exhausted before Second Amendment rights apply?

- Provide examples of violent arms.
- Provide examples of nonviolent arms.
- How do networks, cryptocurrency, apps, and blockchain empower a group?

3. Decision power in the context of the Third Amendment:

   "No soldier shall, in time of peace be quartered in any house, without the consent of the owner, nor in time of war, but in a manner to be prescribed by law."

   - Define "consent."
   - Provide an example of a decision that should only be made by an individual, not the group.
   - Who should have the power to decide what constitutes a time of peace or a time of war?
   - In time of war, who should have the power to decide or prescribe whether soldiers "quarter"?

4. Accountability power in the context of the Fourth Amendment:

   "The right of the people to be secure in their persons, houses, papers, and effects, against unreasonable searches and seizures, shall not be violated, and no warrants shall issue, but upon probable cause, supported by oath or affirmation, and particularly describing the place to be searched, and the persons or things to be seized."

- Define "secure."
- Why are warrants documented in writing?
- Why must warrants "particularly describe" that which is being searched or seized?
- Provide a hypothetical example of when a warrant should not be required for a government search.

Endnotes

Chapter 1: Propaganda

1 Tony Fabijancic, *Bosnia: In the Footsteps of Gavrilo Princip* (Edmonton: University of Alberta Press, 2010), p. XXII.
2 Vladimir Dedijer, *The Road to Sarajevo* (New York: Simon and Schuster, 1966), p.188.
3 T.E. Cone Jr., *History of American Pediatrics* (Boston: Little Brown, 1979).
4 Noel Malcolm, *Bosnia: A Short History* (New York: New York University Press, 1996), pp. 140–141.
5 Misha Glenny, *The Balkans: Nationalism, War, and the Great Powers, 1804–2012* (New York: Penguin Books, 2012).
6 Frank L. Kidner, Maria Bucur, Ralph Matheson, Sally McKee, and Theodore Weeks, *Making Europe: The Story of the West Since 1550* (Boston: Cengage Learning, 2013), p. 757.
7 Denis Dzidic, Marija Ristic, Milka Domanovic, Josip Ivanovic, Edona Peci and Sinisa Jakov Marusic, "Gavrilo Princip: Hero or Villain," *The Guardian*, May 6, 2014, https://www.theguardian.com/world/2014/may/06/gavrilo-princip-hero-villain-first-world-war-balkan-history.

8 Marija Ristic, "Serbia Unveils Monument to Gavrilo Princip," *Balkin Insight*, June 29, 2015, https://balkaninsight.com/2015/06/29/serbia-reveals-monument-to-gavrilo-princip/.

9 Manuela Kasper-Claridge, "Gavrilo Princip, assassin who sparked WWI, gets statue in Belgrade," *Deutsche Welle*, June 29, 2015, https://p.dw.com/p/1Fojx.

10 Jennifer King and Lucy Sweeney, "Archduke Franz Ferdinand: The man whose assassination is blamed for triggering World War," *ABC News*, June 27, 2014, https://www.abc.net.au/news/2014-06-28/franz-ferdinand-profile/5542910.

11 Heeresgeschichtliches Museum Wien, *Wikimedia Commons*, June 3, 2011, https://commons.wikimedia.org/wiki/File:Franz_Ferdinand_Gr%C3%A4f_%2B_Stift_Car.jpg.

12 Tim Butcher, "The lie that started the First World War," *The Telegraph*, June 28, 2014, https://www.telegraph.co.uk/history/world-war-one/10930829/The-lie-that-started-the-First-World-War.html.

13 Ari Shapiro, "A Century Ago In Sarajevo: A Plot, A Farce and a Fateful Shot," *NPR*, June 27, 2014, https://www.npr.org/2014/06/27/325516359/a-century-ago-in-sarajevo-a-plot-a-farce-and-a-fateful-shot.

14 Luigi Albertini, *Origins of the War of 1914* (London: Oxford University Press, 1952–1953).

15 Vladimir Dedijer, *The Road to Sarajevo, World War, 1914–1918* (New York: Simon and Schuster, 1966).

16 John Keegan, *The First World War* (New York: Knopf, 1999).

17 David Fromkin, *Europe's Last Summer: Why the World Went to War in 1914* (New York: Heinemann, 2004).

Chapter 2: False Flags

18 James Weland, "Misguided Intelligence: Japanese Military Intelligence Officers in the Manchurian Incident, September 1931," *The Journal of Military History*, vol. 58, no. 3 (July 1994), pp. 445–460.

19 Robert H. Ferrell, "The Mukden Incident: September 18–19, 1931," *Journal of Modern History* (Chicago: University of Chicago Press, 1955), pp. 66–72.

20 Richard Sims, *Japanese Political History Since the Meiji Renovation 1868–2000* (New York: Palgrave Macmillan, 2001), p. 156.

21 Jamie Bisher, *White Terror: Cossack Warlords of the Trans-Siberian* (New York: Routledge, 2005), p. 299, https://books.google.com/books?id=t8sdihXN47wC&printsec=frontcover&source=gbs_ge_summary_r&cad=0#v=onepage&q=doihara&f=false.

22 Victor Bulwer-Lytton, *League of Nations, Appeal by the Chinese Government, Report by the Commission of Enquiry*, C. 663. M. 320., October 1, 1932, p. 26, https://www.wdl.org/en/item/11601/view/1/25/.

23 Edward Behr, *The Last Emperor* (New York: Bantam Books, 1987), p. 180.

24 James William Morley, *Japan Erupts: The London Naval Conference and the Manchurian Incident, 1928–1932* (New York: Columbia University Press, 1984).

25 Sandra Wilson, *The Manchurian Crisis and Japanese Society, 1931–33* (New York: Routledge, 2002).

26 Mark Felton, *Japan's Gestapo: Murder, Mayhem and Torture in Wartime Asia* (Barnsley, UK: Pen & Sword, 2009).

27 Justus D. Doenecke, *When the Wicked Rise: American Opinion-Makers and the Manchurian Crisis of 1931–1933* (Lewisburg, PA: Bucknell University Press, 1984).

28 Ian Kershaw, *Hitler: 1889–1936 Hubris* (New York: W. W. Norton, 1999).

29 Volker Ullrich, *Hitler: Ascent, 1889–1939* (New York: Vintage Books, 2016).

30 Nazi Regime, "The 'Rohm Putsch' 1934," *Lebendiges Museum Online*, https://www.dhm.de/lemo/kapitel/ns-regime/etablierung/roehm/.
31 David Irving, *Goebbels: Mastermind of the Third Reich* (London: Focal Point, 1996).
32 Roger Manvell and Heinrich Fraenkel, *Heinrich Himmler: The Sinister Life of the Head of the SS and Gestapo* (New York: Skyhorse Publishing, 2007).
33 World War II Casualties, *Wikipedia*, retrieved August 16, 2017, https://en.wikipedia.org/wiki/World_War_II_casualties.

Chapter 3: Censorship

34 Randall L. Bytwerk, *Landmark Speeches of National Socialism* (College Station: Texas A&M University Press, 2008).
35 G.M. Gilbert, *Nuremberg Diary* (Cambridge: Da Capo Press, 1995), p. 278.
36 Rebecca Braun and Lyn Marven, *Cultural Impact in the German Context: Studies in Transmission, Reception, and Influence* (Rochester: Camden House, 2010), p. 177.
37 Seymour Rossel, *The Holocaust: The World and the Jews, 1933–1945* (New Jersey: Behrman House 1992), p. 93.

Chapter 4: Surveillance

38 Scott Shane and Tom Bowman, "Rigging the Game Spy sting: Few at the Swiss factory knew the mysterious visitors were pulling off a stunning intelligence coup—perhaps the most audacious in the National Security Agency's long war on foreign codes," *The Baltimore Sun*, December 10, 1995, https://www.baltimoresun.com/news/bs-xpm-1995-12-10-1995344001-story.html.

39 Greg Miller and Peter F. Mueller, "Compromised encryption machines gave CIA window into major human rights abuses in South America," *The Washington Post*, February 17, 2020, https://www.washingtonpost.com/national-security/compromised-encryption-machines-gave-cia-window-into-major-human-rights-abuses-in-south-america/2020/02/15/bbfa5e56-4f63-11ea-b721-9f4cdc90bc1c_story.html.

40 Richard J. Evans, *The Third Reich in Power 1933–1939* (London: Penguin, 2006), p. 110.

41 William Shirer, *The Rise and Fall of the Third Reich* (New York: Simon & Schuster, 1990).

42 Ian Kershaw, *Hitler, 1889–1936* (London: Penguin, 1998).

43 Edward Bernays, *Propaganda* (New York: Ig Publishing, 1955), p. 48.

44 Adolf Hitler and Ralph Mannheim, *Mein Kampf* (Boston: Houghton-Mifflin, 1999), p. 183.

45 Dennis W. Johnson, *Routledge Handbook of Political Management* (New York: Routledge, 2009).

46 David Irving, *Goebbels: Mastermind of the Third Reich* (London: Focal Point, 1996).

47 Jacques Ellul, *Propaganda: The Formation of Men's Attitudes* (New York: Random House, 1965).

Chapter 5: Secrecy Laws

48 John Pilger, "The lies of Hiroshima are the lies of today," *johnpilger.com*, August 6, 2008, http://johnpilger.com/articles/the-lies-of-hiroshima-are-the-lies-of-today.

49 George F. Will, "One of humanity's remarkable achievements is the absence of the use of a third nuclear weapon," *The Washington Post*, August 5, 2020, https://www.washingtonpost.com/opinions/thankfully-history-has-not-yet-repeated-hiroshima-and-nagasaki/2020/08/04/eb2e4af2-d67c-11ea-930e-d88518c57dcc_story.html.

50 Sarah Pruitt, "The Hiroshima Bombing Didn't Just End WWII—It Kick-Started the Cold War," *history.com*, December 19, 2018, https://www.history.com/news/hiroshima-nagasaki-bombing-wwii-cold-war.
51 Robert S. Burrell, *The Ghosts of Iwo Jima* (College Station: Texas A&M University Press, 2011), p. 118.
52 Oliver Stone and Peter Kuznick, "Bombing Hiroshima changed the world, but it didn't end WWII," *L.A. Times*, May 26, 2016, http://www.latimes.com/opinion/op-ed/la-oe-stone-kuznick-hiroshima-obama-20160524-snap-story.html.
53 John Pilger, "The lies of Hiroshima live on, props in the war crimes of the 20th century," *The Guardian*, August 5, 2008, https://www.theguardian.com/commentisfree/2008/aug/06/secondworldwar.warcrimes.
54 Greg Mitchell, "66 Years Ago: When Truman Opened the Nuclear Era with a Hiroshima Lie," *Huffington Post*, October 6, 2001, https://www.huffingtonpost.com/greg-mitchell/66-years-ago-when-truman_b_920095.html.
55 Pierre Bienaime, "Here's How Many 'Super Nukes' American Scientists Thought It Would Take to Destroy the World in 1945," *Business Insider*, December 16, 2014, https://www.businessinsider.com/heres-how-many-super-nukes-it-would-take-to-destroy-the-world-2014-12?op=1.
56 Kelsey Davenport, "Nuclear Weapons: Who Has What at a Glance," *Arms Control Association*, October 3, 2017, https://www.armscontrol.org/factsheets/Nuclearweaponswhohaswhat.
57 Paul Ham, *Hiroshima Nagasaki: The Real Story of the Atomic Bombings and Their Aftermath* (New York: St. Martin's Press, 2011).
58 "New Mexico Rancher's 'Flying Disk' Proves to be Weather Balloon-Kite," *Fort Worth Star-Telegram*, July 9, 1947.

59 John Dilmore, "Q&A with Luis Elizondo," *Roswell Daily Record*, June 26, 2021, https://www.rdrnews.com/2021/06/26/qa-with-luis-elizondo/.

60 Ancient Aliens, "Beyond Roswell," Season 11, Episode 13, *History Channel*, March 11, 2019, https://www.youtube.com/watch?v=lvimRsH1O_4.

61 Brad Steiger and Donald R. Schmitt, *Project Blue Book: The Top Secret UFO Files that Revealed a Government Cover-Up (MUFON)* (Newburyport: Weiser, 2019).

62 Tim Weiner, *Legacy of Ashes: The History of the CIA* (New York: Random House, 2008).

63 *United States v. Richardson*, 418 U.S. 166 (1974), https://supreme.justia.com/cases/federal/us/418/166/case.html.

64 *50 U.S.C. §401* (National Security Act of 1947), https://www.law.cornell.edu/uscode/text/50/chapter-15.

65 *50 U.S.C. §403a* (Central Intelligence Agency Act), https://law.justia.com/codes/us/1997/title50/chap15/subchapi/sec403a/.

66 Michael A. Ledeen, *Accomplice to Evil: Iran and the War Against the West* (New York: St. Martin's Press, 2009), p. 140.

67 Jim Baggott, *The First War of Physics: The Secret History of the Atom Bomb, 1939–1949* (New York: Pegasus Books, 2010).

68 Jeffrey T. Richelson, "The Secret History of the U-2," *The National Security Archive*, August 15, 2013, https://nsarchive2.gwu.edu/NSAEBB/NSAEBB434/.

69 Richard H. Graham, *SR-71 Revealed: The Inside Story* (St. Paul: MBI Publishing Company, 1996).

70 Gary Shepard, "CIA and Nugan Hand Bank," *CBS News*, September 22, 1982, https://www.youtube.com/watch?v=T0GgF3Z3lSQ.

71 Nicholas Shaxson, *Treasure Islands: Tax Havens and the Men who Stole the World* (New York: Random House, 2011).

72 Annie Jacobsen, *Operation Paperclip: The Secret Intelligence Program that Brought Nazi Scientists to America* (New York: Hatchette Book Group, 2014).

73 Reinhard Gehlen, *The Service: The Memoirs of General Reinhard Gehlen* (New York: Popular Library, 1973).

74 Avalon Project, "Military-Industrial Complex Speech, Dwight D. Eisenhower," *Yale Law School*, January 17, 1961, http://avalon.law.yale.edu/20th_century/eisenhower001.asp.

75 John S. Bowman, *Pergolesi in the Pentagon: Life at the Front Lines of the Cultural Cold War* (Bloomington: Xlibris, 2014), p. 103.

Chapter 6: Mass Media

76 Carl Bernstein, "The CIA and the Media," *Rolling Stone*, https://www.carlbernstein.com/the-cia-and-the-media-rolling-stone-10-20-1977?rq=Cia%2C%20media.

77 John M. Crewdson, "Worldwide Propaganda Network Built by the CIA," *New York Times*, December 26, 1977, https://timesmachine.nytimes.com/timesmachine/1977/12/26/75326395.html?pageNumber=1.

78 Tim Weiner, "F. Mark Wyatt, 86, CIA Officer, Is Dead," *New York Times*, July 6, 2006, http://www.nytimes.com/2006/07/06/us/06wyatt.html?mtrref=www.google.com&gwh=CB8AF3E52848C58F243F1CAEF6BB5485&gwt=pay.

79 William Blum, *Killing Hope: U. S. Military and CIA Interventions Since World War II* (Monroe, ME: Common Courage, 1995), Chapters 2 and 18.

80 Carl Bernstein, "The CIA and the Media," *Rolling Stone*, October 20, 1977, https://www.carlbernstein.com/the-cia-and-the-media-rolling-stone-10-20-1977?rq=Cia%2C%20media.

81 "Alcide de Gasperi," *Time*, April 19, 1948, http://content.time.com/time/covers/0,16641,19480419,00.html.

82 Janet Abu-Lughod, "Israeli Settlements in Occupied Arab Lands: Conquest to Colony," *Journal of Palestine Studies*, vol. 11, issue 2, 1982.
83 Kenneth Pollack, *A Path Out of the Desert: A Grand Strategy for America in the Middle East* (New York: Random House, 2009), p. 30.
84 Adnan Khan, *100 Years of the Middle East: The Struggle for the Post Sykes-Picot Middle East* (Seattle: CreateSpace, 2016), p. 123.
85 Peter Grose, *Continuing the Inquiry: The Council on Foreign Relations from 1921 to 1996* (New York: Council on Foreign Relations Press, 1990), p. 37.
86 James Perloff, *The Shadows of Power: The Council on Foreign Relations and the American Decline* (Appleton, WI: Western Islands, 1988).
87 Daniel Gordis, *Israel: A Concise History of a Nation Reborn* (New York: Collins, 2016).
88 Ari Shavit, *My Promised Land: The Triumph and Tragedy of Israel* (New York: Random House, 2013).
89 Andrej Kreutz, *Russia in the Middle East: Friend or Foe?* (Santa Barbara: Praeger, 2006).
90 Tim Weiner, *Legacy of Ashes: The History of the CIA* (New York: Random House, 2008), p. 160.
91 William Blum, *Killing Hope: U. S. Military and CIA Interventions Since World War II* (Monroe, ME: Common Courage, 1995), Chapter 12.
92 "Conspiracy: In 1957, Syria became the first Arab country to expel an American diplomat, after CIA planned to overthrow the Syrian government," *Parhlo* (Pakistan), October 9, 2014, https://www.parhlo.com/conspiracy-in-1957-syria-became-the-first-arab-country-to-expel-an-american-diplomat-after-cia-planned-to-overthrow-the-syrian-government/.
93 James A. Paul, *Human Rights in Syria* (New York: Human Rights Watch, 1990).

94 Heesu Kim, *Anglo-American Relations and the Attempts to Settle the Korean Question 1953–1960* (London: School of Economics and Political Science, 1996).

95 Robert A. Devine, T.H. Breen, George M. Frederickson, R. Hal Williams, Adriela J. Gross, and H.W. Brands, *America Past and Present, Vol. II: Since 1865* (New York: Pearson Longman, 2007).

96 Michael Pembroke, *Korea: Where the American Century Began* (London: Hardie Grant Books, 2018).

97 "Memorial Hall of the War to Resist U.S. Aggression and Aid Korea is now reopened with new building," *chinamartyrs.gov*, September 21, 2020, http://www.chinamartyrs.gov.cn/x_xwzx/gdyw/202009/t20200921_42206.html?_zbs_baidu_bk.

98 "War to resist U.S. aggression and aid Korea," *China Internet Information Center*, June 27, 2016, http://www.china.org.cn/archive/2016-06/27/content_38756407.htm.

99 William Blum, *Killing Hope: U. S. Military and CIA Interventions Since World War II* (Monroe, ME: Common Courage, 1995), Chapter 5.

100 Tim Weiner, *Legacy of Ashes: The History of the CIA* (New York: Random House, 2008).

101 Frank Holober, *Raiders of the China Coast: CIA Covert Operations During the Korean War* (Annapolis: Naval Institute Press, 1999).

102 Larry Collins, "The CIA Drug Connection Is as Old as the Agency," *New York Times*, December 3, 1993, http://www.nytimes.com/1993/12/03/opinion/03iht-edlarry.html.

103 William Blum, *Killing Hope: U. S. Military and CIA Interventions Since World War II* (Monroe, ME: Common Courage, 1995), Chapter 9.

104 Kermit Roosevelt, *Counter Coup* (New York: McGraw Hill, 1979).

105 Tim Weiner, *Legacy of Ashes: The History of the CIA* (New York: Random House, 2008).

106 Richard Cavendish, "General Batista Returns to Power in Cuba," *History Today*, March 2002, vol. 52 no. 3.
107 Thomas G. Paterson, *Contesting Castro: The United States and the Triumph of the Cuban Revolution* (Oxford: Oxford University Press, 1995).
108 Timothy Alexander Guzman, "Cuba Pre-1959: The Rise and Fall of a U.S. Backed Dictator with Links to the Mob," *Center for Research on Globalization*, July 23, 2015, https://www.globalresearch.ca/cuba-pre-1959-the-rise-and-fall-of-a-u-s-backed-dictator-with-links-to-the-mob/5464738.
109 Norman V. Walbek and Sidney Weintraub, *Conflict, Order, and Peace in the Americas* (Austin: Lyndon B. Johnson School of Public Affairs, 1978).
110 Tim O'Meilia, "Widow of Cuban dictator Batista dies in WPB," *Palm Beach Post*, October 4, 2006.
111 Bruce Shlain and Martin A. Lee, *Acid Dreams, the Complete Social History of LSD: the CIA, the Sixties, and Beyond* (New York: Grove Press, 1985).
112 Michael Ignatieff, "What Did the CIA Do to His Father," *New York Times*, April 1, 2001, https://www.nytimes.com/2001/04/01/magazine/cia-what-did-the-cia-do-to-his-father.html.
113 "A Study of Assassination," *Freedom of Information Act*, Job No. 79-01025A, Box 73, Folder 4, https://nsarchive2.gwu.edu/NSAEBB/NSAEBB4/ciaguat2.html and https://documents2.theblackvault.com/documents/cia/DOC_0000135832.pdf.
114 "Project MKULTRA, The CIA's Program of Research in Behavioral Modification," *Minutes of the Senate Select Committee on Intelligence Hearing on MKULTRA*, August 3, 1977, https://www.andrew.cmu.edu/user/rp3h/lansberry/mkultra.pdf.
115 Harvey M. Weinstein M.D., *Psychiatry and the CIA: Victims of Mind Control* (Arlington: American Psychiatric Publishing, 1990), p. 129.

116 Opinion of Raymond Close (CIA Chief of Station Saudi Arabia), "Nixon and Faisal: If Arabs mistrust America, there's good reason," *New York Times*, December 19, 2002, http://www.nytimes.com/2002/12/19/opinion/nixon-and-faisal-if-arabs-mistrust-america-theres-good-reason.html.

117 John Marks, *The Search for the 'Manchurian Candidate': The CIA and Mind Control* (New York: W.W. Norton, 1991).

118 Alex Constantine, *Virtual Government: CIA Mind Control Operations in America* (Los Angeles: Feral House, 1997).

119 Karl Kahler, "Costa Rica spy kid: My mom was sent to kill Castro and my dad was president of Venezuela," *The Tico Times* (Costa Rica), October 27, 2015, http://www.ticotimes.net/2015/10/27/costa-rica-spy-kid-my-mom-was-sent-to-kill-castro-and-my-dad-was-president-of-venezuela.

120 Andrew Goliszek, *In the Name of Science* (New York: St. Martin's Press, 2003).

121 Tim Weiner, *Legacy of Ashes: The History of the CIA* (New York: Random House, 2008).

122 David Talbot, *The Devil's Chessboard: Allen Dulles, the CIA, and the Rise of America's Secret Government* (New York: HarperCollins, 2015).

123 Walter LaFeber, *America, Russia, and the Cold War, 1945–1996* (New York: McGraw-Hill, 1996).

124 William Blum, *Killing Hope: U. S. Military and CIA Interventions Since World War II* (Monroe, ME: Common Courage, 1995), Chapters 10 & 37.

125 Susanne Jonas, "50 years later, the lessons from Guatemala," *The Progressive*, June 21, 2004, http://progressive.org/op-eds/50-years-later-lessons-guatemala/.

126 William Blum, *Killing Hope: U.S. Military and CIA Interventions Since World War II* (Monroe: Common Courage, 1995), Chapter 14.

127 Ibid, Chapter 4.
128 Catherine A. Traywick, "Shoes, Jewels, and Monets: The Immense Ill-Gotten Wealth of Imelda Marcos," *Foreign Policy*, January 16, 2014, https://foreignpolicy.com/2014/01/16/shoes-jewels-and-monets-the-immense-ill-gotten-wealth-of-imelda-marcos/.
129 David C. Kang, *Crony Capitalism: Corruption and Development in South Korea and the Philippines* (Cambridge: Harvard University Press, 2002).
130 Jo Rodriguez, "10 Conspiracy Theories from the Philippines," *Listverse*, October 19, 2014, https://listverse.com/2014/10/19/10-conspiracy-theories-from-the-philippines/.
131 William Blum, *Killing Hope: U.S. Military and CIA Interventions Since World War II* (Monroe, ME: Common Courage, 1995), Chapter 20.
132 Penelope McMillan, "Ex-Cambodian President Dies in Fullerton," *L.A. Times*, November 18, 1985, http://articles.latimes.com/1985-11-18/news/mn-7294_1_lon-nol-cambodians.
133 Roger Warner, *Shooting at the Moon: The Story of America's Clandestine War in Laos* (South Royalton NH: Steerforth Press, 1996).
134 Joshua Kurlantzick, *A Great Place to Have a War: America in Laos and the Birth of a Military CIA* (New York: Simon & Schuster, 2016).
135 William Blum, *Killing Hope: U.S. Military and CIA Interventions Since World War II* (Monroe, ME: Common Courage, 1995), Chapter 21.
136 Sibel Edmonds, "NATO-CIA-Pentagon: Junction of the Real Druglords and Warlords," *Center for Research on Globalization*, August 22, 2017, http://www.globalresearch.ca/nato-cia-pentagon-junction-of-the-real-druglords-and-warlords/5605177.
137 William Stevenson, *The Revolutionary King: The True-Life Sequel to the 'King and I'* (London: Robinson Publishing, 2001).

138 Paul M. Handley, *The King Never Smiles: A Biography of Thailand's Bhumibol Adulyadej* (New Haven: Yale University Press, 2006).

139 Chris Kohler, "The business of royalty: The five richest monarchs in the world," *The Australian*, April 25, 2014, http://www.theaustralian.com.au/business/business-spectator/the-business-of-royalty-the-five-richest-monarchs-in-the-world/news-story/7c2d2426f7b047f018daf14f799649f9.

140 Nancy B. Tucker, *Patterns in the Dust: Chinese-American Relations and the Recognition Controversy, 1949–1950* (New York: Columbia University Press, 1983).

141 Peter R. Moody, *Opposition and Dissent in Contemporary China* (Washington, D.C.: Hoover Press, 1977).

142 William Blum, *Killing Hope: U.S. Military and CIA Interventions Since World War II* (Monroe, ME: Common Courage, 1995).

143 Bob Woodward, *Veil: The Secret Wars of the CIA, 1981–1987* (New York: Simon & Schuster, 1987).

144 Simon Hooper, "The rise and fall of Noriega, Central America's strongman," *CNN*, July 7, 2010, http://www.cnn.com/2010/WORLD/americas/07/07/panama.manuel.noriega.profile/index.html.

145 Bob Woodward, *Veil: The Secret Wars of the CIA, 1981–1987* (New York: Simon & Schuster, 1987).

146 Julie M. Bunck and Michael R. Fowler, *Bribes, Bullets, and Intimidation: Drug Trafficking and the Law in Central America* (University Park: Penn State University Press, 2012).

147 William Blum, *Killing Hope: U.S. Military and CIA Interventions Since World War II* (Monroe, ME: Common Courage, 1995), Chapter 27.

148 Peter Dale Scott and Jonathan Marshall, *Cocaine Politics: Drugs, Armies, and the CIA in Latin America* (Berkeley: University of California Press, 1998).

149 William Blum, *Killing Hope: U.S. Military and CIA Interventions Since World War II* (Monroe, ME: Common Courage, 1995), Chapter 54.
150 Bob Woodward, *Veil: The Secret Wars of the CIA, 1981–1987* (New York: Simon & Schuster, 1987).
151 John Dinges, *The Condor Years: How Pinochet and His Allies Brought Terrorism to Three Continents* (New York: The New Press, 2005).
152 Peter Dale Scott and Jonathan Marshall, *Cocaine Politics: Drugs, Armies, and the CIA in Latin America* (Berkeley: University of California Press, 1998).
153 William Blum, *Killing Hope: U.S. Military and CIA Interventions Since World War II* (Monroe, ME: Common Courage, 1995), Chapter 36.
154 Stephen Zunes, "US Intervention in Bolivia," *Huffington Post*, October 23, 2008, https://www.huffingtonpost.com/stephen-zunes/us-intervention-in-bolivi_b_127528.html.
155 Dana Priest, "Covert action in Colombia," *Washington Post*, December 21, 2013, http://www.washingtonpost.com/sf/investigative/2013/12/21/covert-action-in-colombia/?utm_term=.cd32f725485e.
156 Peter Dale Scott, *Drugs, Oil, and War: The United States in Afghanistan, Colombia, and Indochina* (Oxford: Rowman and Littlefield, 2003).
157 William Blum, *Killing Hope: U.S. Military and CIA Interventions Since World War II* (Monroe, ME: Common Courage, 1995), Chapter 49.
158 Bob Woodward, *Veil: The Secret Wars of the CIA, 1981–1987* (New York: Simon & Schuster, 1987).
159 John Perkins, *Confessions of an Economic Hit Man* (San Francisco: Berrett-Koehler, 2004).
160 Ebenezer Obiri Addo, *Kwame Nkrumah: A Case Study of Religion and Politics in Ghana* (Lanham, MD: University Press of America, 1999), p. 94.
161 William Blum, *Killing Hope: U.S. Military and CIA Interventions Since World War II* (Monroe, ME: Common Courage, 1995), Chapter 32.

162 Peter Barker, *Operation Cold Chop: The Coup That Toppled Nkrumah* (Accra: Ghana Publishing, 1969).

163 Seymour M. Hersh, "C.I.A. Said to Have Aided Plotters Who Overthrew Nkrumah in Ghana," *New York Times,* May 9, 1978, http://www.nytimes.com/1978/05/09/archives/cia-said-to-have-aided-plotters-who-overthrew-nkrumah-in-ghana.html.

164 Daniele Ganser, *NATO's Secret Armies: Operation GLADIO and Terrorism in Western Europe* (New York: Frank Cass, 2005), see Introduction.

165 "OWSGV: The Austrian stay-behind organisation during the Cold War," *cryptomuseum.com*, December 22, 2016, http://www.cryptomuseum.com/spy/gladio/at/index.htm#ref.

166 Gordon Duff, "Gladio: How We Terrorized Ourselves," *Veterans Today*, September 12, 2012, https://www.veteranstodayarchives.com/2012/09/12/gladio-how-we-terrorized-ourselves-archival/25/.

167 William Blum, *Killing Hope: U.S. Military and CIA Interventions Since World War II* (Monroe, ME: Common Courage, 1995), Chapter 24.

168 "Frenchwomen, Frenchmen," *Life*, vol. 50, no. 18, p. 22, May 5, 1961, https://books.google.com/books?id=qE8EAAAAMBAJ&pg=PA22&lpg=PA22&dq=Frenchwomen,+Frenchmen,+help+me&source=bl&ots=N-DAinZZhMD&sig=Aq14t-YQTCwuxPfgKwt-ub8pt-js&hl=en&sa=X&ved=0ahUKEwi5pOTj8_jXAhUE8m-MKHe9eC88Q6AEITTAI#v=onepage&q=Frenchwomen%2C%20Frenchmen%2C%20help%20me&f=false.

169 "Money: De Gaulle v. the Dollar," *Time*, February 12, 1965, http://content.time.com/time/magazine/article/0,9171,840572,00.html.

170 Daniele Ganser, *NATO's Secret Armies: Operation GLADIO and Terrorism in Western Europe* (New York: Frank Cass, 2005).

171 Franco Ferraresi, *Threats to Democracy: The Radical Right in Italy after the War* (Princeton: Princeton University Press, 1996).
172 Paul L. Williams, *Operation Gladio: The Unholy Alliance between the Vatican, the CIA, and the Mafia* (Westminster: Prometheus Books, 2015).
173 Alvin Shuster, "Ex-Premier Rumor Linked to Lockheed," *New York Times*, December 3, 1976, http://www.nytimes.com/1976/12/03/archives/expremier-rumor-linked-to-lochheed-italian-committee-accuses-him.html.
174 Richard Drake, *The Aldo Moro Murder Case* (Cambridge: Harvard University Press, 1995).
175 Gerhard Feldbauer, *Agenten, Terror, Staatskomplott: Der Mord an Aldo Moro, Rote Brigaden, und CIA* (Cologne: PapyRossa, 2000).
176 Robert D. McFadden, "5 Are Acquitted in Brooklyn of Plot to Run Guns to IRA," *New York Times*, November 6, 1982, http://www.nytimes.com/1982/11/06/nyregion/5-are-acquitted-in-brooklyn-of-plot-to-run-guns-to-ira.html.
177 Douglas Martin, "Michael Flannery, an Advocate of a United Ireland, Dies at 92," *New York Times*, October 2, 1994, http://www.nytimes.com/1994/10/02/obituaries/michael-flannery-an-advocate-of-a-united-ireland-dies-at-92.html.
178 M.A. Birand, *The Generals' Coup in Turkey: An Inside Story of 12 September 1980* (Sterling, VA: Potomac Books, 1987).
179 "How the Turkish government regained control after a failed military coup," *Washington Post*, July 18, 2017, https://www.washingtonpost.com/graphics/world/turkey-military-coup/.
180 Cristina Maza, "Did this CIA Agent try to Overthrow a Foreign Government? Who is Graham Fuller?" *Newsweek*, December 1, 2017, http://www.newsweek.com/cia-graham-fuller-arrest-turkey-erdogan-gulen-dugin-coup-2016-zarrab-728425.

181 "Turkey," U.S. Foreign Aid by Country, *USAID*, retrieved December 1, 2017, https://explorer.usaid.gov/cd/TUR.
182 Daniele Ganser, "The British Secret Service in Neutral Switzerland," *Intelligence and National Security*, December 20, 2005.
183 Felix Wursten, "Conference 'NATO Secret Armies and P26'—The Dark Side of the West," *ETH Zurich Institute*, February 10, 2005.
184 Ralph McGehee, "CIA and the Crisis of Democracy," *CIABASE*, November 14, 2001, reproduced at http://www.serendipity.li/cia/ciabase/ciabase_report_3.htm.
185 Stephen Kinzer, *The Brothers: John Foster Dulles, Allen Dulles, and Their Secret World War* (New York: St. Martin's, 2015), p. 317.
186 Ralph McGehee, *Deadly Deceits: My 25 Years in the CIA* (New York: Open Road, 2002).
187 "Statistical information about casualties of the Vietnam War," *National Archives Military Records*, April 29, 2008, https://www.archives.gov/research/military/vietnam-war/casualty-statistics.html.
188 William Blum, *Rogue State* (Monroe, ME: Common Courage Press, 2005).
189 H. Keith Melton, *CIA Special Weapons & Equipment: Spy Devices of the Cold War* (New York: Sterling Publishing, 1993).
190 Michael Warner, *The CIA Under Harry Truman—CIA Cold War Records* (Honolulu: University Press of the Pacific, 2005).
191 "Alleged Plots Involving Foreign Leaders," *U.S. Senate Select Committee to Study Governmental Operations with Respect to Intelligence Activities*, S. Rep. No. 755, 94th Congress, 2nd Session, https://www.cia.gov/library/readingroom/docs/CIA-RDP83-01042R000200090002-0.pdf.

192 Peter G. Bourne, *Fidel: A Biography of Fidel Castro* (New York City: Dodd Mead & Company, 1986).
193 "Fidel Castro: Dodging exploding seashells, poison pens and ex-lovers," *BBC*, November 27, 2016, https://www.bbc.com/news/world-latin-america-38121583.
194 Anita Snow, "CIA Plot to Kill Castro Detailed," *The Washington Post*, June 27, 2007.
195 "Pentagon Proposed Pretexts for Cuban Invasion in 1962," *The Joint Chiefs of Staff National Security Archive*, March 13, 1962, https://nsarchive2.gwu.edu/news/20010430/northwoods.pdf.
196 Ann Louise Bardach, "The Story of Marita Lorenz: Mistress, Mother, CIA Informant, and Center of Swirling Conspiracy Theories," *Vanity Fair*, November 1993, https://www.vanityfair.com/culture/2016/03/marita-lorenz-fidel-castro-conspiracy-theories.
197 Haynes Johnson, *The Bay of Pigs: The Leaders' Story of Brigade 2506* (New York: W.W. Norton, 1974).
198 William Blum, *Killing Hope: U.S. Military and CIA Interventions Since World War II* (Monroe, ME: Common Courage, 1995), Chapter 30.
199 David M. Barrett, *The CIA and Congress: The Untold Story from Truman to Kennedy* (Lawrence: University Press of Kansas, 2005).
200 Deborah Davis, *Katharine the Great: Katharine Graham and the Washington Post* (San Diego: Harcourt Brace Jovanovich, 1979).
201 Paul David Pope, *The Deeds of My Fathers: How My Grandfather and Father Built New York and Created the Tabloid World of Today* (Lanham: Rowman & Littlefield, 2010), p. 309.
202 Linda Baletsa, *Operation Mockingbird* (Boston: Spratt, 2013).
203 David Talbot, *The Devil's Chessboard: Allen Dulles, the CIA, and the Rise of America's Secret Government* (New York: HarperCollins, 2015).

204 E. Howard Hunt, *American Spy: My Secret History in the CIA, Watergate, and Beyond* (Hoboken: John Wiley & Sons, 2007).

205 Dr. James Tracy, "Interview 12: Robert David Steele," *Real Politik*, August 2, 2014, https://www.youtube.com/watch?v=1W_P6guvN4Y.

206 Frances Stonor Saunders, *The Cultural Cold War: The CIA and the World of Arts and Letters* (New York: The New Press, 2000), p. 105.

207 Richard Helms, *A Look Over My Shoulder: A Life in the Central Intelligence Agency* (New York: Random House, 2003).

208 Nicholas Schou, *Spooked: How the CIA Manipulates the Media and Hoodwinks Hollywood* (New York: Hot Books, 2016).

Chapter 7: Controlling Economies

No endnotes.

Chapter 8: Manipulating Options

209 See https://en.wikipedia.org/wiki/List_of_breakfast_cereals.

210 Bill Allison and Sarah Harkins, "Fixed Fortunes: Biggest corporate political interests spend billions, get trillions," *Sunlight Foundation*, November 17, 2014, https://sunlightfoundation.com/2014/11/17/fixed-fortunes-biggest-corporate-political-interests-spend-billions-get-trillions/.

Chapter 9: Political Election Rigging

211 William Blum, *Killing Hope: U.S. Military and CIA Interventions Since World War II* (Monroe, ME: Common Courage, 1995).

212 "War in Iraq: Not a Humanitarian Intervention," *Human Rights Watch*, January 25, 2004, https://www.hrw.org/news/2004/01/25/war-iraq-not-humanitarian-intervention.

213 Andrew Cockburn and Patrick Cockburn, *Out of the Ashes, The Resurrection of Saddam Hussein* (London: Verso, 2000).

214 Brian Ross, "What's Left of Saddam's Fortune?" *ABC News*, December 3, 2003, http://abcnews.go.com/WNT/Investigation/story?id=129291.

215 "Not My Job: Former CIA Officer Robert Baer Gets Quizzed On Bears," transcript of Peter Sagal interview of former CIA Agent Robert Baer, *NPR*, January 10, 2015, https://www.npr.org/2015/01/10/376096789/not-my-job-former-cia-officer-robert-baer-gets-quizzed-on-bears.

216 "Questionable Activities," *George Washington University National Security Archive*, January 17, 1975, https://nsarchive.gwu.edu/document/20837-1.

217 John Perkins, *Confessions of an Economic Hit Man* (San Francisco: Berrett-Koehler, 2004).

218 Warren Richey, "Noriega Strategy Unfolds Attorneys' Hope to Drag Past U.S. Role Into Trial," *Sun Sentinel*, May 1, 1991, http://articles.sun-sentinel.com/1991-05-01/news/9101220014_1_frank-rubino-gen-noriega-panama.

219 Barbara Tasch, "Former Malaysian Leader Accuses CIA of Cover-Up in Missing Jet," *Time*, May 19, 2014, https://time.com/104480/malaysia-airliens-flight-370-mahathir-mohamad/.

220 Georges Nzongola-Ntalaja, "Patrice Lumumba: the most important assassination of the 20th century," *The Guardian*, January 17, 2011, https://www.theguardian.com/global-development/poverty-matters/2011/jan/17/patrice-lumumba-50th-anniversary-assassination.

221 William Blum, *Killing Hope: U.S. Military and CIA Interventions Since World War II* (Monroe, ME: Common Courage, 1995), Chapters 26 and 42.

222 John Stockwell, *In Search of Enemies: A CIA Story* (New York: W.W. Norton, 1978).

223 Stephen Weissman, "Opening the Secret Files on Lumumba's Murder," *Washington Post*, July 21, 2002, reproduced at http://www1.udel.edu/globalagenda/2003/student/readings/CIAlumumba.html.

224 Emmanuel Gerard and Bruce Kuklick, *Death in the Congo: Murdering Patrice Lumumba* (Cambridge: Harvard University Press, 2015).

225 Harris M. Lentz, "Haiti," in *Heads of States and Governments* (Abingdon, UK: Routledge, 2014).

226 William Blum, *Killing Hope: U.S. Military and CIA Interventions Since World War II* (Monroe, ME: Common Courage, 1995), Chapters 22, 29, and 55.

227 "CIA 'Family Jewels' Memo," *George Washington University National Security Archive*, May 16, 1973, https://nsarchive2.gwu.edu//NSAEBB/NSAEBB222/family_jewels_full_ocr.pdf.

228 Tim Mansel, "I shot the cruelest dictator in the Americas," *BBC News*, May 28, 2011, http://www.bbc.com/news/world-latin-america-13560512.

229 Elizabeth Abbott, *Haiti: A Shattered Nation* (New York: Overlook Duckworth, 2011).

230 "Dominican Republic and Haiti: Country Studies," *Library of Congress Federal Research Division* (Washington, D.C.: Library of Congress, 1999), https://archive.org/stream/dominicanrepubli00metz#page/288/mode/2up.

231 Stephanie Hanes, "Jean-Claude Duvalier, ex-Haitian leader known as Baby Doc, dies at 63," *Washington Post*, October 4, 2014, https://www.washingtonpost.com/world/the_americas/jean-claude-duvalier-ex-haitian-leader-known-as-baby-doc-dies-at-63/2014/10/04/ecdaa2bc-4be3-11e4-b72e-d60a9229cc10_story.html?utm_term=.eda74257a762.

232 "JFK and the Diem Coup," *National Security Archive*, November 5, 2003.

233 David Wallechinsky and Irving Wallace, *The Peoples' Almanac* (New York: Doubleday, 1975).
234 Bui Diem, *In the Jaws of History* (New York: Houghton Mifflin, 1987).
235 Jon Schwarz, "In 1974 Call to Abolish CIA, Sanders Followed in Footsteps of JFK, Truman," *The Intercept*, February 22, 2016, https://theintercept.com/2016/02/22/in-1974-call-to-abolish-cia-sanders-followed-in-footsteps-of-jfk-truman/.
236 David Talbot, *The Devil's Chessboard: Allen Dulles, the CIA, and the Rise of America's Secret Government* (New York: HarperCollins, 2015).
237 Nicholas M. Horrock, "Oswald Link to CIA Reported at Inquiry," *New York Times*, March 27, 1978, http://www.nytimes.com/1978/03/27/archives/oswald-link-to-cia-reported-at-inquiry-exemployee-of-agency-tells.html.
238 Philip Shenon, "Yes, the CIA Director Was Part of the JFK Assassination Cover-Up," *Politico*, October 6, 2015, https://www.politico.com/magazine/story/2015/10/jfk-assassination-john-mccone-warren-commission-cia-213197/.
239 Marita Lorenz, *Marita: The Spy Who Loved Castro* (New York: Pegasus Books, 2017), see "From the Official Story to the Truth."
240 Colonel John Hughes-Wilson, *JFK: An American Coup D'etat: The Truth Behind the Kennedy Assassination* (London: John Blake, 2016).
241 Ann Louise Bardach, "The Story of Marita Lorenz: Mistress, Mother, CIA Informant, and Center of Swirling Conspiracy Theories," *Vanity Fair*, November 1, 1993, https://www.vanityfair.com/culture/2016/03/marita-lorenz-fidel-castro-conspiracy-theories.
242 Katie Bo Lillis, "JFK researchers underwhelmed by latest release of assassination documents," *CNN*, December 15, 2021, https://www.cnn.com/2021/12/15/politics/biden-administration-jfk-documents/index.html.

243 Nicholas M. Horrock, "Oswald Link to CIA Reported at Inquiry," *New York Times*, March 27, 1978, http://www.nytimes.com/1978/03/27/archives/oswald-link-to-cia-reported-at-inquiry-exemployee-of-agency-tells.html.

244 Marita Lorenz, *Marita: The Spy Who Loved Castro* (New York: Pegasus Books, 2017), see "From the Official Story to the Truth."

245 Colonel John Hughes-Wilson, *JFK: An American Coup D'etat: The Truth Behind the Kennedy Assassination* (London: John Blake, 2016).

246 Martin Tolchin, "How Johnson Won Election He'd Lost," *New York Times*, February 11, 1990, http://www.nytimes.com/1990/02/11/us/how-johnson-won-election-he-d-lost.html.

247 Anthony Summers, *Official and Confidential: The Secret Life of J. Edgar Hoover* (New York: Open Road Media, 2012).

248 "The Assassination of John F. Kennedy, the Clint Murchison Meeting," *YouTube*, https://www.youtube.com/watch?v=POmdd6HQsus.

249 Seb Menard, "JFK Murder Confession by CIA Agent," *YouTube*, https://www.youtube.com/watch?v=PWCYanma0YM.

250 Erik Hedegaard, "The Last Confession of E. Howard Hunt," *Rolling Stone*, April 5, 2007, https://www.rollingstone.com/feature/the-last-confession-of-e-howard-hunt-76611/.

251 Antonio Veciana, Carlos Harrison, and David Talbot, *Trained to Kill: The Inside Story of CIA Plots against Castro, Kennedy, and Che* (New York: Skyhorse, 2017).

252 "JFK Assassination System," *National Archives*, https://www.archives.gov/files/research/jfk/releases/docid-32263509.pdf.

253 Stephen M. Underhill, *The Manufacture of Consent: J. Edgar Hoover and the Rhetorical Rise of the FBI* (East Lansing: Michigan State University Press, 2020).

254 "JFK Assassination: Former FBI Agent Don Adams on Joseph Milteer," *YouTube*, https://www.youtube.com/watch?v=6vjrc7hdcYk.
255 Philip Deane, *I Should Have Died* (New York: Atheneum, 1977), pp. 113–114.
256 George Crile, *Charlie Wilson's War* (New York: Grove Press, 2003), p. 52.
257 David Heymann Clemens, *American Legacy: The Story of John and Caroline Kennedy*, (New York: Simon and Schuster, 2007).

Chapter 10: White Power

258 Benjamin Karim, *Remembering Malcolm* (New York: Carroll & Graf, 1992).
259 Louis E. Lomax, *When the Word Is Given: A Report on Elijah Muhammad, Malcolm X, and the Black Muslim World* (Cleveland: World Publishing, 1963).
260 Alex Haley, *The Autobiography of Malcolm X* (New York: Ishi Press, 1965).
261 Bruce Perry, *Malcolm: The Life of a Man Who Changed Black America* (New York: Station Hill, 1991).
262 Manning Marable, *Malcolm X: A Life of Reinvention* (New York: Viking, 2011).
263 Peter Kihss, "Malcolm X Shot to Death at Rally Here," *New York Times*, February 22, 1965.
264 Karl Evanzz, *The Judas Factor: The Plot to Kill Malcolm X* (New York: Basic Books, 1992).
265 Leonard J. Moore, *Citizen Klansmen: The Ku Klux Klan in Indiana, 1921–1928* (Chapel Hill: University of North Carolina Press, 1997).
266 Mabry v. State, 110 So. 2d 250 (Ala. Ct. App. 1959), https://www.courtlistener.com/opinion/1847346/mabry-v-state/.
267 Glenn T. Eskew, *But for Birmingham: The Local and National Movements in the Civil Rights Struggle* (Chapel Hill: University of North Carolina Press, 1997).

268 Edward Harris, *Miracle in Birmingham: A Civil Rights Memoir, 1954–1965* (New York: Stonework Press, 2004).

269 Stephen B. Oates, *Let the Trumpet Sound: A Life of Martin Luther King, Jr.* (New York: Harper Collins, 1983).

270 John Christoffersen, "MLK Was Inspired by Time in Connecticut," *NBC*, January 17, 2011, https://www.nbcconnecticut.com/news/local/mlk-was-inspired-by-time-in-connecticut/1885278/.

271 Martin L. King Jr., "I Have a Dream" Speech, Lincoln Memorial, Washington, D.C., August 28, 1963, https://kinginstitute.stanford.edu/king-papers/documents/i-have-dream-address-delivered-march-washington-jobs-and-freedom.

272 Mark Engler and Paul Engler, "Why Martin Luther King Didn't Run for President," *Rolling Stone*, January 18, 2016, http://www.rollingstone.com/politics/news/why-martin-luther-king-didnt-run-for-president-20160118.

273 Representative Cynthia McKinney (D-GA), quoted by Jerry Ray and Tamara Carter, "Jerry Ray Declares his Brother James Earl Ray Didn't Shoot Martin Luther King, Jr.," *PRWeb*, January 9, 2012, http://www.prweb.com/releases/2012/1/prweb9075036.htm.

274 Jack E. White, "King Conspiracy Update," *Time*, April 14, 1997, http://content.time.com/time/magazine/article/0,9171,986205,00.html.

275 Jonathan David Farley, "Preventing the rise of a 'messiah,'" *The Guardian*, April 4, 2008, https://www.theguardian.com/commentisfree/2008/apr/04/preventingtheriseofamessi.

276 William F. Pepper, *The Plot to Kill King: The Truth Behind the Assassination of Martin Luther King Jr.* (New York: Skyhorse, 2016).

277 "Complete Transcript of the Martin Luther King, Jr. Assassination Conspiracy Trial," *The King Center*, December 8, 1999, http://www.thekingcenter.org/sites/default/files/KING%20FAMILY%20TRIAL%20TRANSCRIPT.pdf.

278 Robert F. Kennedy, "University of Capetown" Speech, Capetown, South Africa, June 6, 1966, https://www.jfklibrary.org/Research/Research-Aids/Ready-Reference/RFK-Speeches/Day-of-Affirmation-Address-as-delivered.aspx.

279 James DiEugenio, Lisa Pease, and Joe Brown, *The Assassinations: Probe Magazine on JFK, MLK, RFK, and Malcolm X* (Port Townsend: Feral House, 2003).

280 Larry Tye, *Bobby Kennedy: The Making of a Liberal Icon* (New York: Random House, 2016).

281 Shane O'Sullivan, *Who Killed Bobby? The Unsolved Murder of Robert F. Kennedy* (New York: Union Square Press, 2008).

Chapter 11: Double Agents

282 Hersh, Seymour M., *The Price of Power: Kissinger in the Nixon White House* (New York: Summit Books, 1983), chapter 15.

283 Chileng Pa and Carol A. Mortland, *Escaping the Khmer Rouge: A Cambodian Memoir* (New York: McFarland, 2008).

284 Ben Kiernan, *How Pol Pot Came to Power: Colonialism, Nationalism, and Communism in Cambodia, 1930–1975* (New Haven: Yale University Press, 2004).

285 Kyla Ryan, "Cambodia's Ongoing Human Trafficking Problem," *The Diplomat*, July 28, 2014, https://thediplomat.com/2014/07/cambodias-ongoing-human-trafficking-problem/.

286 "The CIA and Chile: Anatomy of an Assassination," *George Washington University National Security Archive*, October 22, 2020, https://nsarchive.gwu.edu/briefing-book/chile/2020-10-22/cia-chile-anatomy-assassination.

287 "CIA Role in Australian Hit," *The Washington Post*, May 5, 1977, http://jfk.hood.edu/Collection/Weisberg%20Subject%20Index%20Files/C%20Disk/CIA%20Australia/Item%2006.pdf.

288 "CIA Triumph," *The Village Voice*, February 27, 1976, http://jfk.hood.edu/Collection/White%20Materials/Security-CIA/CIA%201481.pdf.

289 John Pilger, "The British-American coup that ended Australian independence," *The Guardian*, October 23, 2014, https://www.theguardian.com/commentisfree/2014/oct/23/gough-whitlam-1975-coup-ended-australian-independence.

290 William Blum, *Killing Hope: U.S. Military and CIA Interventions Since World War II* (Monroe, ME: Common Courage, 1995), chapter 40.

291 "Crash Mrs. Hunt Died in Blamed on Pilot Error," *St. Petersburg Times*, September 28, 1973, https://news.google.com/newspapers?id=oOpRAAAAIBAJ&pg=2905%2C2476956.

292 Erik Hedegaard, "The Last Confession of E. Howard Hunt," *Rolling Stone*, April 5, 2007, https://www.rollingstone.com/feature/the-last-confession-of-e-howard-hunt-76611/.

293 "Colson's Weird Scenario," *Time*, July 8, 1974, http://content.time.com/time/subscriber/article/0,33009,943904-1,00.html.

294 "Aircraft Accident Report, United Airlines Inc., Boeing 737, N9031U Chicago-Midway Airport, Chicago, Illinois, December 8, 1972," NTSB-AAR-73-16, *National Transportation Safety Board*, August 29, 1973.

295 "Chilean president Salvador Allende committed suicide, autopsy confirms," *The Guardian*, July 19, 2011, https://www.theguardian.com/world/2011/jul/20/salvador-allende-committed-suicide-autopsy.

296 William Blum, *Killing Hope: U.S. Military and CIA Interventions Since World War II* (Monroe, ME: Common Courage, 1995), chapter 34.

297 John Dinges, *The Condor Years: How Pinochet and His Allies Brought Terrorism to Three Continents* (New York: The New Press, 2005).
298 Peter Kornbluh, "Chile and the United States: Declassified Documents Relating to the Military Coup, September 11, 1973," *George Washington University National Security Archive*, September 11, 1998, https://nsarchive2.gwu.edu//NSAEBB/NSAEBB8/nsaebb8.htm.
299 "Life under Pinochet: 'They were taking turns to electrocute us one after the other,'" *Amnesty International*, September 11, 2013, https://www.amnesty.org/en/latest/news/2013/09/life-under-pinochet-they-were-taking-turns-electrocute-us-one-after-other/.
300 Daurius Figueira, *Cocaine and Heroin Trafficking in the Caribbean* (New York: iUniverse, 2004).
301 Gary Webb, *The Dark Alliance* (New York: Seven Stories Press, 2014).
302 Casey Gane-McCalla, "Jamaica's Shower Posse: How the CIA Created 'the Most Notorious Criminal Organization,'" *Centre for Global Research*, June 13, 2010, https://www.globalresearch.ca/jamaica-s-shower-posse-how-the-cia-created-the-most-notorious-criminal-organization/19696.
303 Laurie Gunst, *Born fi' Dead* (New York: Henry Holt, 1995).
304 Casey Gane-McCalla, *Inside the CIA's Secret War in Jamaica* (Los Angeles: Over the Edge Books, 2016).
305 Don Taylor, *Marley and Me: The Real Bob Marley Story* (Fort Lee, NJ: Barricade Books, 1995).
306 Daurius Figueira, *Cocaine and Heroin Trafficking in the Caribbean* (New York: iUniverse, 2004).
307 "Saudi Arabia: The Death of a Desert Monarch," *Time*, April 7, 1975, http://content.time.com/time/magazine/article/0,9171,917224-2,00.html#ixzz1AyvreNfM.

308 Raymond Close, "Nixon and Faisal: If Arabs mistrust America, there's good reason," *New York Times*, December 19, 2002, http://www.nytimes.com/2002/12/19/opinion/nixon-and-faisal-if-arabs-mistrust-america-theres-good-reason.html.

309 Andrew Tully, *CIA: The Inside Story* (Lake Oswego, OR: eNet Press, 2012).

Chapter 12: Intimidation

310 Jorge Ferreira, *Joao Goulart* (Rio De Janeiro: Civilizacao Brasileira, 2011), p. 110.

311 Robert D. McFadden, "Lincoln Gordon Dies at 96; Educator and Ambassador to Brazil," *New York Times*. December 21, 2009, https://www.nytimes.com/2009/12/21/us/21GORDON.html?_r=1.

312 Peter Kornbluh, "Brazil Marks 40th Anniversary of Military Coup," *The National Security Archive*, March 31, 2004, https://nsarchive2.gwu.edu/NSAEBB/NSAEBB118/index.htm.

313 James G. Hershberg and Peter Kornbluh, "Brazil Marks 50th Anniversary of Military Coup," *National Security Archive*, April 2, 2014, https://nsarchive2.gwu.edu/NSAEBB/NSAEBB465/.

314 Eric Konigsberg, "Who Killed Anna Mae?" *New York Times*, April 25, 2014.

315 Heidi Bell Gease, "Witness testifies FBI agent threatened Aquash's Life," *Rapid City Journal*, December 3, 2010, https://rapidcityjournal.com/news/witness-testifies-fbi-agent-threatened-aquashs-life/article_0292eb72-ff36-11df-8d22-001cc4c03286.html.

316 Johanna Brand, *The Life and Death of Anna Mae Aquash* (Toronto: James Lorimer, 1993).

317 Peter Matthiessen, *In the Spirit of Crazy Horse: The Story of Leonard Peltier and the FBI's War on the American Indian Movement* (New York: Penguin Books, 1992).

318 Patrice McSherry, *Predatory States: Operation Condor and Covert War in Latin America* (Lanham, MD: Rowman & Littlefield, 2005).
319 Duncan Campbell, "Kissinger Approved Argentinian 'Dirty War,'" *The Guardian*, December 6, 2003, https://www.theguardian.com/world/2003/dec/06/argentina.usa.
320 Daniel A. Grech, "Transcript: U.S. OK'd 'Dirty War,'" *Miami Herald*, December 4, 2003, https://nsarchive2.gwu.edu/NSAEBB/NSAEBB104/herald.pdf.
321 Marcela Valente, "Argentina: Remains of Mothers of Plaza de Mayo Identified," *Inter Press Service*, July 8, 2005, http://www.ipsnews.net/2005/07/argentina-remains-of-mothers-of-plaza-de-mayo-identified.
322 Matthew Wills, "The Stolen Children of Argentina," *JSTOR Daily*, August 22, 2018, https://daily.jstor.org/stolen-children-of-argentina.
323 Calvin Sims, "Argentine Tells of Dumping 'Dirty War' Captives into Sea," *New York Times*, March 13, 1995, https://www.nytimes.com/1995/03/13/world/argentine-tells-of-dumping-dirty-war-captives-into-sea.html?pagewanted=all.
324 Jonathan Mann, "Macabre new details emerge about Argentina's 'Dirty War,'" *CNN*, March 23, 1996, http://www.cnn.com/WORLD/9603/argentina.war/index.html.
325 Nabi Misdaq, *Afghanistan: Political Frailty and External Interference* (Abingdon: Taylor & Francis, 2006).
326 William Blum, *Killing Hope: U.S. Military and CIA Interventions Since World War II* (Monroe, ME: Common Courage, 1995).
327 Barnett R. Rubin, *The Fragmentation of Afghanistan* (New Haven: Yale University Press, 2002).
328 Steve Coll, *Ghost Wars: The Secret History of the CIA, Afghanistan, and Bin Laden, from the Soviet Invasion to September 10, 2001* (New York: Penguin Books, 2004).

329 Donald L. Barlett and James B. Steele, "The Oily Americans," *Time*, May 13, 2003, https://web.archive.org/web/20081204024027/http://www.time.com/time/magazine/article/0,9171,450997-92,00.html.
330 Sani H. Panhwar, "CIA Sent Bhutto to the Gallows," *New York Times*, April 5, 1979.
331 Mohammad Yousaf, *Silent Soldier: The Man behind the Afghan Jehad General Akhtar Abdur Rahman* (Karachi, Sindh: Jang Publishers, 1991).
332 Piero Gleijeses, "The Case for Power Sharing in El Salvador," *Foreign Affairs: Council on Foreign Relations, 1983*, vol. 61, no. 5, https://www.jstor.org/stable/20041635.
333 *Mesoamerica* (Costa Rica: Institute for Central American Studies, 1982).
334 Belisaric Betancur, Reinaldo Figueredo Planchart, and Thomas Buergenthal, "Report of the United Nations Truth Commission on El Salvador: The 12-year war in El Salvador," *United Nations Security Council*, April 1, 1993, http://www.derechos.org/nizkor/salvador/informes/truth.html.
335 Tracy L. Barnett, "The four churchwomen murdered in El Salvador," *Global Sisters Report*, June 19, 2017, https://www.globalsistersreport.org/news/four-churchwomen-murdered-el-salvador-47386.
336 Tracy Wilkinson, "US-trained Atlacatl unit was famed for battle prowess but was also implicated in atrocities," *L.A. Times*, December 9, 1992, https://www.latimes.com/archives/la-xpm-1992-12-09-mn-1714-story.html.
337 Webster G. Tarpley and Anton Chaitkin, *George Bush: The Unauthorized Biography* (Joshua Tree: Progressive Press, 2004).
338 Russ Baker, *Family of Secrets: The Bush Dynasty, America's Invisible Government, and the Hidden History of the Last Fifty Years* (New York: Bloomsbury Press, 2009).

339 Russ Baker, "Bush Angle to Reagan Shooting Still Unresolved as Hinckley Walks," *WhoWhatWhy.org*, August 16, 2016, https://whowhatwhy.org/2016/08/16/bush-angle-reagan-shooting-still-unresolved-hinckley-walks/.

340 Ronald J. Ostrow, "Webster Chosen as CIA Director: President Picks FBI Chief to Head Agency Under Fire for Iran Role," *L.A. Times*, March 4, 1987, http://articles.latimes.com/1987-03-04/news/mn-4592_1_fbi-director-william-h-webster.

341 Ari Shapiro, "Database Tracks History of U.S. Meddling in Foreign Elections," *NPR*, December 22, 2016, https://www.npr.org/2016/12/22/506625913/database-tracks-history-of-u-s-meddling-in-foreign-elections.

342 Mark Weisbrot, "Hard choices: Hillary Clinton admits role in Honduran coup aftermath," *Al Jazeera*, September 29, 2014, http://america.aljazeera.com/opinions/2014/9/hillary-clinton-honduraslatinamericaforeignpolicy.html.

343 Stephen Zunes, "The U.S. Role in The Honduras Coup and Subsequent Violence," *Huffington Post*, June 19, 2016, https://www.huffingtonpost.com/entry/the-us-role-in-the-honduras-coup-and-subsequent-violence_us_5766c7ebe4b0092652d7a138.

344 James Hodge and Linda Cooper, "The CIA and Abu Ghraib: 50 Years of Teaching and Training Torturers," *Counterpunch*, November 3, 2004, reproduced at https://web.archive.org/web/20091003205201/http://www.counterpunch.org/hodge11032004.html.

345 Eric Zuesse, "Head of Stratfor, 'Private CIA,' Says Overthrow of Yanukovych Was 'the Most Blatant Coup in History,'" *Washingtonsblog.com*, December 20, 2014, http://www.washingtonsblog.com/2014/12/head-stratfor-private-cia-says-overthrow-yanukovych-blatant-coup-history.html. See also https://www.globalresearch.ca/head-of-stratfor-private-cia-says-overthrow-of-yanukovych-was-the-most-blatant-coup-in-history/5420978?pdf=5420978.

346 "Congressional Record—House of Representatives," *Government Publishing Office*, September 3, 1997, https://www.gpo.gov/fdsys/pkg/GPO-CRECB-1997-pt12/pdf/GPO-CRECB-1997-pt12-5-2.pdf.

347 Nicolas J.S. Davies, "35 countries where the U.S. has supported fascists, drug lords and terrorists," *Salon*, March 8, 2014, https://www.salon.com/2014/03/08/35_countries_the_u_s_has_backed_international_crime_partner/.

348 Senator Daniel Inouye, "Iran-Contra Hearings Closing Statement," *Senate Select Committee on Secret Military Assistance to Iran and the Nicaraguan Opposition*, August 3, 1987, reproduced at http://www.danielkinouyeinstitute.org/speeches/iran-contra-hearings-closing-statement.

Chapter 13: Financial and Other Incentives

349 Jonathan D. Moreno, "Harvard's Experiment on the Unabomber, Class of '62," *Psychology Today*, May 25, 2012, https://www.psychologytoday.com/us/blog/impromptu-man/201205/harvards-experiment-the-unabomber-class-62.

350 "MKUltra: Inside the CIA's Cold War mind control experiments," *The Week*, July 21, 2017, http://www.theweek.co.uk/86961/mkultra-inside-the-cias-cold-war-mind-control-experiments.

351 Colin A. Ross, *The CIA Doctors* (Richardson: Manitou Communications, 2006).

352 Carol M. Ostrom, "Unabomber Suspect Is Charged—Montana Townsfolk Showed Tolerance for 'The Hermit,'" *The Seattle Times*, April 6, 1996, http://community.seattletimes.nwsource.com/archive/?date=19960404&slug=2322396.

353 Karl Stampfl, "He came Ted Kaczynski, he left The Unabomber," *The Michigan Daily*, March 16, 2006, https://web.archive.org/web/20170114062259/https://www.michigandaily.com/content/he-came-ted-kaczynski-he-left-unabomber.

354 Robert D. McFadden, "Prisoner of Rage… From a Child of Promise to the Unabom Suspect," *New York Times*, May 26, 1996, https://www.nytimes.com/1996/05/26/us/prisoner-of-rage-a-special-report-from-a-child-of-promise-to-the-unabom-suspect.html.

355 Alston Chase, "Harvard and the Making of the Unabomber," *The Atlantic*, June 1, 2000, https://www.theatlantic.com/magazine/archive/2000/06/harvard-and-the-making-of-the-unabomber/378239/.

356 Jonathan D. Moreno, "Harvard's Experiment on the Unabomber, Class of '62," *Psychology Today*, May 25, 2012, https://www.psychologytoday.com/blog/impromptu-man/201205/harvards-experiment-the-unabomber-class-62.

357 Gregory Korte, "The rise of Rick Singer: How the mastermind of college admissions scandal built an empire on lies, exploited a broken system," *USA Today*, June 19, 2019, https://www.usatoday.com/in-depth/news/nation/2019/06/19/college-admission-bribery-scandal-rick-singer-exploited-broken-system-loughlin-huffman/1133729001/.

358 "'Dark Waters' star Mark Ruffalo, lawyer Rob Bilott tell the true story behind film," *Nightline*, November 26, 2019, https://www.youtube.com/watch?v=Tkkuil-U6qQ.

359 Nicole Wendee, "PFOA and Cancer in a Highly Exposed Community: New Findings from the C8 Science Panel," *Environmental Health Perspectives*, Nov-Dec 2013, https://www.ncbi.nlm.nih.gov/pmc/articles/PMC3855507/.

Chapter 14: Conflicts of Interest

360 Kenneth B. Noble, "Tailhook Whistle-Blower Recalls Attack," *New York Times*, October 4, 1994, https://www.nytimes.com/1994/10/04/us/tailhook-whistle-blower-recalls-attack.html.

361 Ibid.

362 Katherine Boo, "Universal soldier: what Paula Coughlin can teach American women—sexual assault victim demands justice," *Washington Monthly*, September 1992.

363 David Knowles, "From Tailhook Whistleblower to Warrior Pose," *Wall Street Journal*, April 30, 2009, https://web.archive.org/web/20141025035138/http://magazine.wsj.com/hunter/second-chapter/from-navy-whistleblower-to-warrior-pose/.

364 Jessica Guynn, "Sexual harassment used to cost women their careers…," *USA Today*, December 4, 2017, https://www.usatoday.com/story/money/business/2017/12/04/sexual-harassment-women-careers/901611001/.

365 Ralph Franklin Keeling, *Gruesome Harvest: The Allies' Postwar War against the German People* (Torrance: Institute for Historical Review, 1992).

366 Aleksandr I. Solzhenitsyn, *The Gulag Archipelago, 1918–1956: An Experiment in Literary Investigation* (New York: Harper & Row, 1974).

367 Keith Lowe, *Savage Continent: Europe in the Aftermath of World War II* (New York: St. Martin's Press, 2012).

368 Letters, "Army Behavior," *Time*, vol. XLVI no. 12, September 17, 1945, http://content.time.com/time/subscriber/article/0,33009,854433-2,00.html.

369 Kevin Alfred Strom, "Ravishing the Women of Conquered Europe," *Christian Century*, December 5, 1945, http://www.library.flawlesslogic.com/massrape.htm.

370 Kim Kozlowski, "How MSU Doc Became Suspect in Dozens of Rapes," *The Detroit News*, August 10, 2017, https://www.detroitnews.com/story/news/local/michigan/2017/08/10/rise-fall-larry-nassar/104491508/.

371 Alice Park, "Why It's a Big Deal That USA Gymnastics Is Cutting Ties with the Karolyi Ranch," *Time*, January 18, 2018, https://time.com/5108887/usa-gymnastics-sexual-abuse-karolyi-ranch/.

372 "Who is Larry Nassar? A timeline of his decades-long career, sexual assault convictions and prison sentences," *USA Today*, August 22, 2018, https://www.usatoday.com/pages/interactives/larry-nassar-timeline/.

373 Juana Macias, "Athlete A," *Netflix*, June 24, 2020, https://www.netflix.com/title/81034185.

374 Erica Tempesta and Charlie Lankston, "Larry Nassar victim Maggie Nichols opens up about first time he assaulted her—as her parents claim USAG punished her for reporting pedophile doctor's abuse by leaving her off the 2016 Olympic team," *Daily Mail*, June 24, 2020, https://www.dailymail.co.uk/femail/article-8455477/Did-USAG-punish-Maggie-Nichols-reporting-Larry-Nassar-abuse.html.

375 John Barr, "USA Gymnastics struck deal with McKayla Maroney to keep Larry Nassar abuse quiet, lawyer says," *ESPN*, December 20, 2017, https://www.espn.com/olympics/gymnastics/story/_/id/21825575/usa-gymnastics-struck-agreement-mckayla-maroney-keep-larry-nassar-abuse-quiet-lawyer-says.

376 Tim Evans, Tony Cook, Marisa Kwiatkowski, and Sarah Bowman, "Indianapolis FBI leader eyed head USA Gymnastics job after sitting on Nassar allegations," *Indianapolis Star*, July 16, 2021, https://www.indystar.com/story/news/investigations/2021/07/16/larry-nassar-usa-gymnastics-fbi-investigation-doj-job-report/7977052002/.

377 Alice Park, "Who Is Larry Nassar, the Former USA Gymnastics Doctor Aly Raisman Accuses of Sexual Abuse?" *Time*, October 18, 2017, http://time.com/4988116/larry-nassar-mckayla-maroney-sexual-abuse-doctor-usa-gymnastics/.

378 "Untold: Operation Flagrant Foul," *Netflix*, September 1, 2022, https://www.netflix.com/title/81026443.

379 Tom Goldman, "Ex-Referee Says 2002 NBA Playoff Was Rigged," *NPR*, June 12, 2008, https://www.npr.org/2008/06/12/91415111/ex-referee-says-2002-nba-playoff-was-rigged.

380 "Untold: Operation Flagrant Foul," *Netflix*, September 1, 2022, https://www.netflix.com/title/81026443.

Chapter 15: Quid Pro Quo

381 Ronan Farrow, "From Aggressive Overtures to Sexual Assault: Harvey Weinstein's Accusers Tell Their Stories," *The New Yorker*, October 10, 2017, https://www.newyorker.com/news/news-desk/from-aggressive-overtures-to-sexual-assault-harvey-weinsteins-accusers-tell-their-stories.

382 Jodi Kantor and Megan Twohey, "Harvey Weinstein Paid Off Sexual Harassment Accusers for Decades," *New York Times*, October 5, 2017, https://www.nytimes.com/2017/10/05/us/harvey-weinstein-harassment-allegations.html.

383 Ronan Farrow, "Harvey Weinstein's Army of Spies," *The New Yorker*, November 6, 2017, https://www.newyorker.com/news/news-desk/harvey-weinsteins-army-of-spies.

384 Lesley Goldberg, "Roy Price Resigns as Amazon Studios Head," *Hollywood Reporter*, October 17, 2017, https://www.hollywoodreporter.com/tv/tv-news/roy-price-resigns-as-amazon-studios-head-1049728/.

385 Philip Shabecoff, "Global Warming Has Begun, Expert Tells Senate," *New York Times*, June 24, 1988, http://www.nytimes.com/1988/06/24/us/global-warming-has-begun-expert-tells-senate.html?pagewanted=all.

386 Chris McGreal, "Big oil and gas kept a dirty secret for decades. Now they may pay the price," *The Guardian*, June 30, 2021, https://www.theguardian.com/environment/2021/jun/30/climate-crimes-oil-and-gas-environment.

387 Spencer R. Weart, *The Discovery of Global Warming* (Cambridge: Harvard University Press, 2008).

388 Emily Holden, "Oil and gas industry rewards U.S. lawmakers who oppose environmental protections—study," *The Guardian*, February 24, 2020, https://www.theguardian.com/environment/2020/feb/24/oil-gas-industry-us-lawmakers-campaign-donations-analysis.

389 Bob Fitrakis and Harvey Wasserman, "New Hampshire the Birthplace of Electronic Election Theft," *freepress.org*, February 8, 2016, https://freepress.org/article/new-hampshire-birthplace-electronic-election-theft.

390 Melvin A. Goodman, *Whistleblower at the CIA: An Insider's Account of the Politics of Intelligence* (San Francisco: City Lights Books, 2017), p. 188.

391 Edward Mickolus, *Stories from Langley: A Glimpse Inside the CIA* (Lincoln: University of Nebraska Press, 2014), p. 84.

Chapter 16: Criminalization

392 Douglas Valentine, "Creating a Crime: How the CIA Commandeered the DEA," *Counterpunch*, September 11, 2015, https://www.counterpunch.org/2015/09/11/creating-a-crime-how-the-cia-commandeered-the-dea/.

393 21 U.S.C. §801 (Controlled Substances Act).

394 Russ Baker, *Family of Secrets: The Bush Dynasty, America's Invisible Government, and the Hidden History of the Last Fifty Years* (New York: Bloomsbury Publishing, 2009).

395 Douglas Valentine, "Creating a Crime: How the CIA Commandeered the DEA," *Counterpunch*, September 11, 2015, https://www.counterpunch.org/2015/09/11/creating-a-crime-how-the-cia-commandeered-the-dea/.

396 Howard Kohn, "Cowboy in the Capital: Drug Czar Bill Bennett," *Rolling Stone*, November 2, 1989, http://www.rollingstone.com/culture/features/cowboy-in-the-capital-drug-czar-bill-bennett-19891102.

397 George H.W. Bush, "Address to the Nation on the National Drug Control Strategy," *American Presidency Project*, September 5, 1989, https://www.presidency.ucsb.edu/documents/address-the-nation-the-national-drug-control-strategy.

398 Douglas Valentine, "Creating a Crime: How the CIA Commandeered the DEA," *Counterpunch*, September 11, 2015, https://www.counterpunch.org/2015/09/11/creating-a-crime-how-the-cia-commandeered-the-dea/.

399 Tom LoBianco, "Report: Aide says Nixon's war on drugs targeted blacks, hippies," *CNN*, March 24, 2016, http://www.cnn.com/2016/03/23/politics/john-ehrlichman-richard-nixon-drug-war-blacks-hippie/index.html.

400 Adam Andrzejewski, "War Weapons for America's Police Departments: New Data Shows Feds Transfer $2.2B In Military Gear," *Forbes*, May 10, 2016, https://www.forbes.com/sites/adamandrzejewski/2016/05/10/war-weapons-for-americas-local-police-departments/#163380ac4af4.

401 Donna Selman and Paul Leighton, *Punishment for Sale: Private Prisons, Big Business, and the Incarceration Binge* (Lanham, MD: Rowman & Littlefield, 2010).

402 "War Comes Home: The Excessive Militarization of American Policing," *American Civil Liberties Union*, June 2014, https://www.aclu.org/report/war-comes-home-excessive-militarization-american-police.

403 Eric Schlosser, "The Prison-Industrial Complex," *The Atlantic*, December 1998, https://www.theatlantic.com/magazine/archive/1998/12/the-prison-industrial-complex/304669/.

404 Douglas Valentine, *The CIA as Organized Crime: How Illegal Operations Corrupt America and the World* (Atlanta: Clarity Press, 2017).

405 Gary Webb, *Dark Alliance: The CIA, the Contras, and the Crack Cocaine Explosion* (New York: Seven Stories Press, 1999).

406 Mike Sager, "Say Hello to Rick Ross," *Esquire*, September 25, 2013, http://www.esquire.com/news-politics/a25818/rick-ross-drug-dealer-interview-1013/.

Chapter 17: Violence

407 Rodney King and Lawrence J. Spagnola, *The Riot Within: My Journey from Rebellion to Redemption* (New York: HarperOne, 2012).

408 Frances Kai-Hwa Wang, "Who Is Vincent Chin? The History and Relevance of a 1982 Killing," *NBC News*, June 15, 2017, https://www.nbcnews.com/news/asian-america/who-vincent-chin-history-relevance-1982-killing-n771291.

409 Ibid.

410 William Blum, *Killing Hope: U.S. Military and CIA Interventions Since World War II* (Monroe, ME: Common Courage, 1995), Chapter 53.

411 *U.S. v. Usama Bin Laden*, United States District Court for the Southern District of New York, February 14, 2001, reproduced at http://cryptome.org/usa-v-ubl-05.htm.

412 Steve Coll, *Ghost Wars: The Secret History of the CIA, Afghanistan, and Bin Laden, from the Soviet Invasion to September 10, 2001* (New York: Penguin Books, 2004).

413 "Translation of Bin Laden's Videotaped Message," Transcript, *Washington Post*, November 1, 2004, http://www.washingtonpost.com/wp-dyn/articles/A16990-2004Nov1.html.

414 Michael Moore, "Six Years Ago, Chuck Hagel Told the Truth About Iraq," *Huffington Post*, January 5, 2013, https://www.huffingtonpost.com/michael-moore/chuck-hagel-iraq-oil_b_2414862.html.

415 Peter Beaumont and Joanna Walters, "Greenspan admits Iraq was about oil, as deaths put at 1.2m," *The Guardian*, September 16, 2007, https://www.theguardian.com/world/2007/sep/16/iraq.iraqtimeline.

416 Matt Corley, "Abizaid: 'We've Treated the Arab World as a Collection of Big Gas Stations,'" *ThinkProgress*, October 15, 2007, https://thinkprogress.org/abizaid-weve-treated-the-arab-world-as-a-collection-of-big-gas-stations-cb2612fcd0bd/.

417 Glenn Greenwald, "David Frum, the Iraq war and oil," *The Guardian,* March 18, 2013, https://www.theguardian.com/commentisfree/2013/mar/18/david-frum-iraq-war-oil.

418 "Bin Laden Claims Responsibility for 9/11," *Fox News*, October 30, 2004, http://www.foxnews.com/story/2004/10/30/bin-laden-claims-responsibility-for-11.html.

419 "Translation of Bin Laden's Videotaped Message," *Washington Post*, November 1, 2004, http://www.washingtonpost.com/wp-dyn/articles/A16990-2004Nov1.html.

420 George W. Bush, "Statement by President George W. Bush before the 56th regular session," *U.N. General Assembly*, November 10, 2001, https://www.state.gov/documents/organization/18967.pdf.

421 "Impeaching George W. Bush, President of the United States, of high crimes and misdemeanors," House Resolution 1258, *110th Congress, 2nd Session*, June 11, 2008.

422 Cynthia A. Parker, *Master of Electricity—Nikola Tesla: A Quick-Read Biography About the Life and Inventions of a Visionary Genius* (Seattle: CreateSpace, 2015).

423 William J. Broad, "A Battle to Preserve a Visionary's Bold Failure," *New York Times*, May 4, 2009, http://www.nytimes.com/2009/05/05/science/05tesla.html?mtrref=en.wikipedia.org&gwh=39A6948BD08C281678D42FC246671C0F&gwt=pay.

424 Susan Carew, "Lockheed Skunk Works Director on Deathbed UFO's Are Real," *worldpeacefull.com*, September 4, 2017, https://wpas.worldpeacefull.com/2017/09/lockheed-skunk-works-director-on-deathbed-ufos-are-real/.

425 Jan Harzan, "MUFON," *YouTube*, August 29, 2013, https://www.youtube.com/watch?v=u9ZZekWMiUQ.
426 Robert Alan Goldberg, *Enemies Within: The Culture of Conspiracy in Modern America* (New Haven: Yale University Press, 2008).
427 "Operation Majestic-12 Preliminary Briefing for President-Elect Eisenhower," *bibliotecapleyades.net*, November 18, 1952, reproduced at http://www.bibliotecapleyades.net/imagenes_sociopol/majestic1219_01.gif.
428 Bill Wickersham, "UFOs, Extraterrestrials and Higher Education," *UFO Casebook*, May 7, 2013, http://www.ufocasebook.com/2013/wickersham1.html.
429 Robert Alan Goldberg, *Enemies Within: The Culture of Conspiracy in Modern America* (New Haven: Yale University Press, 2008).
430 Bill Wickersham, "UFO accounts are journalists' stuff of dreams," *Columbia Daily Tribune*, May 27, 2014, https://www.columbiatribune.com/story/opinion/editorials/2014/05/27/ufo-accounts-are-journalists-stuff/985695007/.
431 Pat Paterson, "The Truth About Tonkin," *U.S. Naval Institute*, February 2008, https://www.usni.org/magazines/navalhistory/2008-02/truth-about-tonkin.
432 Neil Sheehan, *The Pentagon Papers: The Secret History of the Vietnam War* (New York: Bantam Books, 1971).
433 Daniel Ellsberg, *Secrets: A Memoir of Vietnam and the Pentagon Papers* (New York: Penguin Books, 2002).

Chapter 18: Coconspirators

434 "Judge William Byrne; Ended Trial Over Pentagon Papers," *Washington Post*, January 15, 2006, http://www.washingtonpost.com/wp-dyn/content/article/2006/01/14/AR2006011401165.html.
435 Carl Bernstein and Bob Woodward, *All the President's Men* (New York: Simon & Schuster, 1974).

436 "William M. Byrne Jr., 75, Judge in the Ellsberg Leak Case, Dies," *New York Times*, January 15, 2006, http://www.nytimes.com/2006/01/15/us/william-m-byrne-jr-75-judge-in-the-ellsberg-leak-case-dies.html.

437 Ibid.

438 Robert H. Bork, *Saving Justice: Watergate, the Saturday Night Massacre, and Other Adventures of a Solicitor General* (New York: Encounter Books, 2013).

439 G. Gordon Liddy, *Will: The Autobiography of G. Gordon Liddy* (New York: St. Martin's Press, 1980).

440 Carroll Kilpatrick, "Nixon Forces Firing of Cox; Richardson, Ruckelshaus Quit," *Washington Post*, October 21, 1973, https://www.washingtonpost.com/wp-srv/national/longterm/watergate/articles/102173-2.htm.

441 Kenneth B. Noble, "Bork Irked by Emphasis on His Role in Watergate," *New York Times*, July 2, 1987, http://www.nytimes.com/1987/07/02/us/bork-irked-by-emphasis-on-his-role-in-watergate.html.

442 "Bork: Nixon Offered Next High Court Vacancy in '73," *Associated Press*, February 25, 2013, https://www.covnews.com/nationworld/bork-nixon-offered-next-high-court-vacancy-in-73/.

443 "*JW v CIA Watergate* CIA report 00146," *Judicial Watch*, August 29, 2016, https://www.judicialwatch.org/wp-content/uploads/2016/08/JW-v-CIA-Watergate-CIA-report-00146.pdf.

444 John Dean, "Nixon's Uses, Abuses and Muses on the Supreme Court," *Verdict*, July 25, 2014, https://verdict.justia.com/2014/07/25/nixons-uses-abuses-muses-supreme-court.

445 Karen DeYoung and Walter Pincus, "CIA to Air Decades of Its Dirty Laundry," *Washington Post*, June 22, 2007, http://www.washingtonpost.com/wp-dyn/content/article/2007/06/21/AR2007062102434.html.

446 Fred Emery, *Watergate* (New York: Simon & Schuster, 1994).

447 Bob Woodward, "How Mark Felt Became 'Deep Throat'," *Washington Post*, June 20, 2005, https://www.washingtonpost.com/politics/how-mark-felt-became-deep-throat/2012/06/04/gJQAlpARIV_story.html?utm_term=.3e23b03d3472.

448 *Buckley v. Valeo*, 424 U.S. 1 (1976), https://supreme.justia.com/cases/federal/us/424/1/case.html.

449 "History of FOIA," *Electronic Frontier Foundation*, https://www.eff.org/issues/transparency/history-of-foia.

450 "Donald Rumsfeld handwritten notes of September 30, 1974," White House staff meeting, *National Security Archive*, https://nsarchive.gwu.edu/document/18212-national-security-archive-doc-03-donald-rumsfeld.

451 "The Select Committee to Study Governmental Operations with Respect to Intelligence Activities," Church Committee report, *U.S. Senate Select Committee on Intelligence* (Washington, D.C.: United States Congress, 1976).

452 "Intelligence Activities and the Rights of Americans," Church Committee report, *U.S. Senate Select Committee on Intelligence* (Washington, D.C.: United States Congress, 1976).

453 Jonny Tickle, "Boris Yeltsin had entourage of 'hundreds' of CIA agents who instructed him how to run Russia, claims former parliamentary speaker," *Russia Today*, June 12, 2021, https://www.rt.com/russia/526345-yeltsin-cia-connection-claim/.

454 Ian Bremmer and Samuel Charap, "The Siloviki in Putin's Russia: Who They Are and What They Want," *The Washington Quarterly*, Center for Strategic and International Studies and the Massachusetts Institute of Technology, Winter 2006-07, https://ciaotest.cc.columbia.edu/olj/twq/win2006-07/07winter_bremmer.pdf.

455 Richard Sakwa, *Putin: Russia's choice* (New York: Routledge, 2008).

456 David Hoffman, "Putin's Career Rooted in Russia's KGB," *The Washington Post*, January 30, 2000, https://www.washingtonpost.com/wp-srv/inatl/longterm/russiagov/putin.htm.

457 Josh Meyer, "A port city, a steel cage, a palace: The steps that made Putin 'the richest man in the world,'" *Yahoo News*, April 27, 2022, https://www.yahoo.com/gma/port-city-steel-cage-palace-070110487.html.

458 Vladimir Kovalev, "Uproar at Honor for Putin," *The Saint Petersburg Times*, July 23, 2004, https://web.archive.org/web/20150320150048/http://sptimes.ru/index.php?action_id=2&story_id=1124.

459 Marshall I. Goldman, "Putin and the Oligarchs," *Council on Foreign Relations*, November 2004, https://web.archive.org/web/20150630005237/http://www.cfr.org/world/putin-oligarchs/p7517.

460 John Witherow, "The odd couple and their extraordinary labyrinth of wealth," *The Times*, January 21, 2012, https://www.thetimes.co.uk/article/the-odd-couple-and-their-extraordinary-labyrinth-of-wealth-3c7qsrvgfbw.

461 David E. Hoffman, *The Oligarchs: Wealth and Power in the New Russia* (New York: Perseus, 2002).

462 Markar Melkonian, "U.S. Meddling in 1996 Russian Elections in Support of Boris Yeltsin," *globalresearch.ca*, November 11, 2017, https://www.globalresearch.ca/us-meddling-in-1996-russian-elections-in-support-of-boris-yeltsin/5568288.

463 "Yanks to the Rescue: The Secret Story of How American Advisers Helped Yeltsin Win," *Time*, vol. 148, no. 4, July 15, 1996.

464 Alessandra Stanley, "To Win Russia's 'Generation X,' Yeltsin Is Pumping Up the Volume," *New York Times*, June 6, 1996, http://www.nytimes.com/1996/06/06/world/to-win-russia-s-generation-x-yeltsin-is-pumping-up-the-volume.html.

465 Oliver Stone, "The Putin Interviews," *Showtime*, 2017, ttps://www.sho.com/the-putin-interviews.
466 Valerie Strauss, "Russia's plagiarism problem: Even Putin has done it," *Washington Post*, March 18, 2014, https://www.washingtonpost.com/news/answer-sheet/wp/2014/03/18/russias-plagiarism-problem-even-putin-has-done-it/.
467 Yuri Felshtinsky and Alexander Litvinenko, *Blowing Up Russia: Terror from Within* (London: Gibson Square Books, 2007).
468 Masha Gessen, *The Man Without a Face: The Unlikely Rise of Vladimir Putin* (London: Granta, 2012).
469 Jamie Grierson, "Litvinenko inquiry: Russia involved in spy's death, Scotland Yard says," *The Guardian*, July 30, 2015, https://www.theguardian.com/world/2015/jul/30/litvinenko-inquiry-russia-involved-spy-death-scotland-yard.
470 Julia Ioffe, "How State-Sponsored Blackmail Works in Russia," *The Atlantic*, January 11, 2017, https://www.theatlantic.com/international/archive/2017/01/kompromat-trump-dossier/512891/.
471 Steven Rosefielde and Stefan Hedlund, *Russia Since 1980* (Cambridge, UK: Cambridge University Press, 2009).
472 Erklärung von Chodorkowski, "Mein besonderer Dank gilt Hans-Dietrich Genscher," *Der Spiegel*, December 20, 2013, https://www.spiegel.de/politik/ausland/erklaerung-von-chodorkowski-dank-an-hans-dietrich-genscher-a-940414.html.
473 Marshall I. Goldman, "Putin and the Oligarchs," *Council on Foreign Relations*, November 2004, https://web.archive.org/web/20150630005237/http://www.cfr.org/world/putin-oligarchs/p7517.
474 Andrew Osborn and Tom Balmforth, "Putin approves changes allowing him to stay in power until 2036," *Reuters*, March 10, 2020, https://www.reuters.com/article/us-russia-politics-idUSKBN20X1FD.

475 Fred Emery, *Watergate: The Corruption of American Politics and the Fall of Richard Nixon* (New York: Simon & Schuster, 1994).
476 92 Stat. 1824, Ethics in Government Act of 1978, *U.S. Congress*, https://www.congress.gov/95/statute/STATUTE-92/STATUTE-92-Pg1824.pdf.
477 Loch K. Johnson, *A Season of Inquiry Revisited: The Church Committee Confronts America's Spy Agencies* (Lawrence: University Press of Kansas, 2015).
478 Linda Greenhouse, "Blank Check; Ethics in Government: The Price of Good Intentions," *New York Times*, February 1, 1998, http://www.nytimes.com/1998/02/01/weekinreview/blank-check-ethics-in-government-the-price-of-good-intentions.html?scp=7&sq=ethics%20in%20government%20act&st=nyt.

Chapter 19: Freedom Fighters
479 Sarah Pruitt, "The Post World War II Boom: How America Got into Gear," *Arts & Entertainment Network*, May 14, 2020, https://www.history.com/news/post-world-war-ii-boom-economy.
480 John Perkins, *Confessions of an Economic Hit Man* (San Francisco: Berrett-Koehler, 2004).
481 Andrew Langley, *September 11: Attack on America* (Mankato, MN: Compass Point Books, 2006).
482 National Commission on Terrorist Attacks Upon the United States, "The Attack Looms," *9/11 Commission*, August 21, 2004, https://govinfo.library.unt.edu/911/report/911Report_Ch7.htm.
483 National Commission on Terrorist Attacks, *The 9/11 Commission Report: Final Report of the National Commission on Terrorist Attacks Upon the United States* (New York: Cosimo Books, 2010).
484 Assaf Moghadam, *The Globalization of Martyrdom: Al Qaeda, Salafi Jihad, and the Diffusion of Suicide Attacks* (Baltimore: Johns Hopkins University Press, 2008).

485 Matthew J. Morgan, *The Impact of 9/11 on Politics and War: The Day that Changed Everything?* (London: Palgrave Macmillan, 2009).

486 Donald L. Barlett and James B. Steele, "The Oily Americans," *Time*, May 13, 2003, https://web.archive.org/web/20081204024027/http://www.time.com/time/magazine/article/0,9171,450997-92,00.html.

487 Bruce Riedel, *What We Won: America's Secret War in Afghanistan, 1979–1989* (Washington D.C.: Brookings Institution Press, 2014).

488 "NSA Whistleblowers William (Bill) Binney and J. Kirk Wiebe," *Government Accountability Project*, December 12, 2013, https://web.archive.org/web/20131212065859/http://www.whistleblower.org/program-areas/homeland-security-a-human-rights/surveillance/nsa-whistleblowers-bill-binney-a-j-kirk-wiebe.

489 Ellen Nakashima, Greg Miller, and Julie Tate, "Former NSA executive Thomas A. Drake may pay high price for media leak," *Washington Post*, July 14, 2010, https://www.washingtonpost.com/wp-dyn/content/article/2010/07/13/AR2010071305992.html.

490 Nick Schwellenbach, "POGO Obtains Pentagon Inspector General Report Associated with NSA Whistleblower Tom Drake," *Project on Government Oversight*, June 22, 2011, https://pogoblog.typepad.com/pogo/2011/06/pogo-obtains-pentagon-inspector-general-report-associated-with-nsa-whistleblower-tom-drake.html.

491 James Risen, "Snowden Says He Took No Secret Files to Russia," *New York Times*, October 17, 2013, http://www.nytimes.com/2013/10/18/world/snowden-says-he-took-no-secret-files-to-russia.html?mtrref=www.google.com&gwh=1955FF58E3FB5200FF647DAE4DE1CEDB&gwt=pay.

492 A.J. Plus, "Exclusive: Edward Snowden on the man who inspired his work," *Al Jazeera*, August 5, 2015, http://america.aljazeera.com/articles/2015/8/5/exclusive-edward-snowden-on-the-man-who-inspired-his-work.html.

493 Julian Assange, *Cypherpunks: Freedom and the Future of the Internet* (New York: OR Books, 2016).

494 Glenn Greenwald, *No Place to Hide: Edward Snowden, the NSA, and the U.S. Surveillance State* (New York: Picador, 2015).

Chapter 20: Self-Regulation

495 Sibel Edmonds, *Boilingfrogs.com*, February 1, 2011, reproduced at http://www.boilingfrogspost.com/tag/boiling-frogs-post-exclusive/feed/. See also https://en.wikipedia.org/wiki/Sibel_Edmonds.

496 Gordon Duff, "Into the Dark: America's Descent into a Police State," *Salem News*, October 3, 2009, http://www.salem-news.com/articles/october032009/police_state_gd_10-3-09.php.

497 *Radack v. U.S. Dept. of Justice*, Memorandum Opinion and Order, August 9, 2005, 402 F. Supp. 99.

498 Eric Lichtblau, *Bush's Law: The Remaking of American Justice* (New York: Pantheon Books, 2008).

499 "A Review of the FBI's Actions in Connection with Allegations Raised by Contract Linguist Sibel Edmonds, Special Report," *Office of the Inspector General, Office of Oversight and Review*, January 2005, https://oig.justice.gov/special/0501/final.pdf.

500 Jesslyn Raddack, *Traitor: The Whistleblower and the "American Taliban"* (Washington, D.C.: Whistleblower Press, 2012).

501 Ellen Nakashima, "Legal memos released on Bush-era justification for warrantless wiretapping," *Washington Post*, September 6, 2014, https://www.washingtonpost.com/world/national-security/legal-memos-released-on-bush-era-justification-for-warrantless-wiretapping/2014/09/05/91b86c52-356d-11e4-9e92-0899b306bbea_story.html?utm_term=.90ba61f9a2f9.

502 Julian Sanchez, "What the Ashcroft 'Hospital Showdown' on NSA spying was all about," *Ars Technica*, July 29, 2013, https://arstechnica.com/tech-policy/2013/07/what-the-ashcroft-hospital-showdown-on-nsa-spying-was-all-about/.

503 Ari Shapiro, "E-Mails Show Justice Dept in Damage-Control Mode," *NPR*, March 20, 2007, https://www.npr.org/templates/story/story.php?storyId=9003390.

Chapter 21: Cover-ups

504 Joseph C. Wilson, "What I Didn't Find in Africa," *New York Times*, July 6, 2003, http://www.nytimes.com/2003/07/06/opinion/what-i-didn-t-find-in-africa.html.

505 Nicholas Rufford, "Italian spies 'faked documents' on Saddam nuclear purchase," *The Sunday Times of London*, August 1, 2004.

506 Amy Goodman, "The Man Who Sold the Iraq War: John Rendon, Bush's General in the Propaganda War," *Democracy Now!*, November 21, 2005, https://www.democracynow.org/2005/11/21/the_man_who_sold_the_iraq.

507 Kevin Liptak, "Trump pardons ex-Cheney aide Scooter Libby," *CNN*, April 13, 2018, https://www.cnn.com/2018/04/13/politics/donald-trump-pardons-scooter-libby/index.html.

508 E. J. Dionne Jr., "How Cheney Fooled Himself," *Washington Post*, June 21, 2005, http://www.washingtonpost.com/wp-dyn/content/article/2005/06/20/AR2005062001177.html.

509 "Kay: No evidence Iraq stockpiled WMDs," *CNN*, January 26, 2004, http://www.cnn.com/2004/WORLD/meast/01/25/sprj.nirq.kay/.

510 "Encore Presentation: Dead Wrong," transcripts, *CNN*, February 25, 2006, http://transcripts.cnn.com/TRANSCRIPTS/0602/25/cp.01.html.

511 Harry Kawilarang, *Quotations on Terrorism* (Victoria: Trafford, 2004), p. 456.

512 Phil Snow (30:49), "Pat Tillman: A Football Life," *NFL Films*, November 6, 2016, http://www.azcardinals.com/videos/videos/Pat-Tillman-A-Football-Life/4e943724-6938-4adb-9eac-cf465df52e61.

513 Michael I. Niman, "Who Killed Pat Tillman?" *Mediastudy.com*, November 10, 2005, http://mediastudy.com/articles/av11-10-05.html.

514 Associated Press, "Tillman Probe Reveals Startling Details," *CBS News*, November 9, 2006, https://www.cbsnews.com/news/tillman-probe-reveals-startling-details/.

515 Marco Margaritoff, "How Did Pat Tillman Die?...," *ATI*, July 8, 2022, https://allthatsinteresting.com/pat-tillman-death.

516 Kyle Dalton, "Evidence Shows Pat Tillman Murdered According to Medical Experts," *Sportscasting*, April 22, 2020, https://www.sportscasting.com/evidence-shows-pat-tillman-murdered-according-to-medical-experts/.

517 Mick Brown, "Betrayal of an all-American hero," *The Telegraph*, October 7, 2010, http://www.telegraph.co.uk/culture/8046658/Betrayal-of-an-all-American-hero.html.

518 "Countdown Thursday: Gonzo with the Wind," Countdown with Keith Olbermann, *MSNBC*, July 26, 2007, http://web.archive.org/web/20071103110131/http://thenewshole.msnbc.msn.com:80/archive/2007/07/26/293623.aspx.

519 Martha Mendoza, "AP: New Details on Tillman's Death," *USA Today*, July 27, 2007, http://web.archive.org/web/20090525150300/http://www.usatoday.com:80/news/nation/2007-07-26-tillman-friendly-fire_N.htm.

520 "Review of Matters Related to the Death of Corporal Patrick Tillman," *Inspector General Report, U.S. Department of Defense*, March 26, 2007, reproduced at https://www.npr.org/documents/2007/mar/tillman/tillman_dod_ig.pdf.

521 Scott Lindlaw and Martha Mendoza, "Army remains uncertain how Pat Tillman died," *Quad-City Times*, November 10, 2006, http://qctimes.com/news/local/army-remains-uncertain-how-pat-tillman-died/article_a59eaa4f-f10c-5c07-910f-46129d5c642f.html.

522 Josh White, "Army Withheld Details About Tillman's Death: Investigator Quickly Learned 'Friendly Fire' Killed Athlete," *Washington Post*, May 4, 2005, http://www.washingtonpost.com/wp-dyn/content/article/2005/05/03/AR2005050301502.html.

523 Matt Renner, "White House 'Stonewalling' on Pat Tillman Documents," *Truth-Out.org*, July 16, 2007, http://truth-out.org/archive/component/k2/item/71885:white-house-stonewalling-on-pat-tillman-documents.

524 Amy Goodman, "Fired Army Whistleblower Receives $970K for Exposing Halliburton No-Bid Contract in Iraq," *Democracy Now!*, July 26, 2001, https://www.democracynow.org/2011/7/26/exclusive_fired_army_whistleblower_receives_970k.

525 Scott Conroy, "Iraq Whistleblowers Vilified, Demoted," *CBS News*, August 25, 2007, https://www.cbsnews.com/news/iraq-whistleblowers-vilified-demoted/.

526 Antonio Taguba, “Taguba Report: AR 15-6 Investigation of the 800th Military Police Brigade,” *American Civil Liberties Union*, October 19, 2004, https://www.thetorturedatabase.org/document/ar-15-6-investigation-800th-military-police-investigating-officer-mg-antonio-taguba-taguba-.

527 Spencer Ackerman, “No looking back: the CIA torture report’s aftermath,” *The Guardian*, September 11, 2016, https://www.theguardian.com/us-news/2016/sep/11/cia-torture-report-aftermath-daniel-jones-senate-investigation.

528 Wesley Bruer, “‘No penalty for CIA employees accused of spying on Senate,” *CNN*, January 14, 2015, http://www.cnn.com/2015/01/14/politics/cia-senate-spying-penalty/index.html.

529 Seymour M. Hersh, “The General’s Report,” *New Yorker*, June 25, 2007, https://www.newyorker.com/magazine/2007/06/25/the-generals-report.

530 Amy Goodman, “Abu Ghraib Whistleblower Samuel Provance Speaks Out on Torture and Cover-Up,” *Democracy Now!*, January 25, 2008, https://www.democracynow.org/2008/1/25/broadcast_exclusive_abu_ghraib_whistleblower_samuel.

531 Chris Hayes, “Russell Tice Transcript,” *NBC News*, June 10, 2013, http://www.nbcnews.com/id/52168305/#.Uurb6_ZiBFQ.

532 Keith Olbermann “Russell Tice—NSA Domestic Spying Targeted Journalists,” *MSNBC*, January 21, 2009, reproduced at https://www.youtube.com/watch?v=NDVJiAFVdZ8.

533 Amy Goodman, “National Security Agency Whistleblower Warns Domestic Spying Program Is Sign the U.S. is Decaying Into a ‘Police State,’” *Democracy Now!*, January 3, 2006, https://www.democracynow.org/2006/1/3/exclusive_national_security_agency_whistleblower_warns.

534 Scott Shane, "A Zealous Watchman to Follow the Money," *New York Times*, March 9, 2009, http://www.nytimes.com/2009/03/10/us/politics/10devaney.html?mtrref=www.google.com&gwh=5B74EB36C96DB-F57BFDA60C17D994AE3&gwt=pay.

535 "Minerals Management Service: False Claims Allegations," *U.S. Department of the Interior, Office of Inspector General*, September 19, 2007, https://www.doioig.gov/sites/doioig.gov/files/Qui-tam.pdf.

536 Charlie Savage, "Sex, Drug Use and Graft Cited in Interior Department," *New York Times*, September 11, 2008, http://www.nytimes.com/2008/09/11/washington/11royalty.html?mtrref=www.google.com&gwh=027A182CC1C98973EBFAF6D4500DDFC3&gwt=pay.

537 Amy Goodman, "Gulf Oil Spill: BP Execs Escape Punishment as Fallout from Disaster Continues to Impact Sea Life," *Democracy Now!*, April 23, 2012, https://www.democracynow.org/2012/4/23/gulf_oil_spill_bp_execs_escape.

Chapter 22: Fraud

538 "HEARING BEFORE THE COMMITTEE ON OVERSIGHT AND GOVERNMENT REFORM HOUSE OF REPRESENTATIVES ONE HUNDRED TENTH CONGRESS FIRST SESSION," October 2, 2007, https://www.archives.gov/congress/hearings.html.

539 Evan Wright, *How to Get Away with Murder in America* (San Francisco: Byliner, 2012).

540 Conor Friedersdorf, "The Terrifying Background of the Man Who Ran a CIA Assassination Unit," *The Atlantic,* July 18, 2012, https://www.theatlantic.com/politics/archive/2012/07/the-terrifying-background-of-the-man-who-ran-a-cia-assassination-unit/259856/.

541 James Risen and Timothy Williams, "US Looks for Blackwater Replacement in Iraq," *New York Times*, January 30, 2009, http://www.nytimes.com/2009/01/30/world/middleeast/30blackwater.html?mtrref=en.wikipedia.org&gwh=70828FAF784809E60AA224757C832B2A&gwt=pay.

542 Jason Ukman, "Ex-Blackwater firm gets a name change, again," *Washington Post*, December 12, 2011, https://www.washingtonpost.com/blogs/checkpoint-washington/post/ex-blackwater-firm-gets-a-name-change-again/2011/12/12/gIQAXf4YpO_blog.html?utm_term=.d6fe1be5125c.

543 Michael Safi and Joshua Robertson, "Australian mercenary reportedly killed in Yemen clashes," *The Guardian*, December 8, 2015, https://www.theguardian.com/australia-news/2015/dec/09/australian-mercenary-reportedly-killed-yemen-clashes.

544 Ibid.

545 Ibid.

546 Kit O'Connell, "Rumors persist that the CIA helps export opium from Afghanistan," *MintPress News*, September 19, 2015, https://www.mintpressnews.com/rumors-persist-that-the-cia-helps-export-opium-from-afghanistan/209687/.

547 Allan MacLeod, "Geopolitics, profit, and poppies: how the CIA turned Afghanistan into a failed narco-state," *MintPress News*, June 25, 2021, https://www.mintpressnews.com/cia-afghanistan-drug-trade-opium/277780/.

548 "Today's Heroin Epidemic," *Center for Disease Control and Prevention*, July 7, 2015, https://www.cdc.gov/vitalsigns/heroin/index.html.

549 Lee Fang, "How Spy Agency Contractors Have Already Abused Their Power," *The Nation*, June 11, 2013, https://www.thenation.com/article/how-spy-agency-contractors-have-already-abused-their-power/.

550 Amy Goodman, "WikiLeaks: Leaked Emails Expose Inner Workings of Private Intelligence Firm Stratfor, a 'Shadow CIA,'" *Democracy Now!*, February 28, 2012, https://www.democracynow.org/2012/2/28/wikileaks_leaked_emails_expose_inner_workings.

551 Dana Milbank, "Tragicomedy of Errors Fuels Volusia Recount," *Washington Post*, November 12, 2000, https://www.washingtonpost.com/archive/politics/2000/11/12/tragicomedy-of-errors-fuels-volusia-recount/5a74f0e0-565b-4980-8ed6-8bed5ff6b19f/?utm_term=.23754cc09c55.

552 Jake Tapper, "Still Some Bugs in Electronic Voting," *ABC News*, March 5, 2004, http://abcnews.go.com/GMA/story?id=127988.

553 Alison Mitchell, "Over Some Objections, Congress Certifies Electoral Vote," *New York Times*, January 7, 2001, http://www.nytimes.com/2001/01/07/us/over-some-objections-congress-certifies-electoral-vote.html.

554 Kim Zetter, "Sequoia Voting Systems Responsible for 2000 Presidential Debacle?" *Wired*, August 15, 2007, https://www.wired.com/2007/08/sequoia-voting/.

555 "Whistleblowers: Sequoia Voting Systems," *National Election Defense Coalition*, retrieved December 7, 2017, reproduced at https://webcache.googleusercontent.com/search?q=cache:hU1cYKrP38sJ:https://www.electiondefense.org/.whistleblowers/+&cd=5&hl=en&ct=clnk&gl=us.

556 "Help America Vote Act," *U.S. Election Assistance Commission*, retrieved November 30, 2017, https://www.eac.gov/about/help-america-vote-act/.

557 "Pew's Electionline.org Examines First Five Years of the Help America Vote Act," *Pew*, November 29, 2007, http://www.pewtrusts.org/en/about/news-room/press-releases/2007/11/29/pews-electionlineorg-examines-first-five-years-of-the-help-america-vote-act.

558 Victoria Collier, "How to Rig an Election," *Harper's*, November 2012, https://harpers.org/archive/2012/11/how-to-rig-an-election/6/.

559 Kim Zetter, "Did E-Vote Firm Patch Election?" *Wired*, October 13, 2003, https://www.wired.com/2003/10/did-e-vote-firm-patch-election/.

560 David M. Halbfinger, "The 2002 Election: Georgia; Bush's Push, Eager Volunteers and Big Turnout Led to Georgia Sweep," *New York Times*, November 10, 2002, http://www.nytimes.com/2002/11/10/us/2002-election-georgia-bush-s-push-eager-volunteers-big-turnout-led-georgia-sweep.html.

561 Peter Soby Jr., "Whistleblower Charged with Three Felonies for Exposing Diebold's Crimes," *Huffington Post*, February 27, 2006, https://www.huffingtonpost.com/peter-soby-jr/whistleblower-charged-wit_b_16411.html.

562 Associated Press, "California Official Seeks Criminal Probe of Evoting," *NBC News*, April 30, 2004, http://www.nbcnews.com/id/4874190/ns/politics-voting_problems/t/california-official-seeks-criminal-probe-e-voting/#.WjFrc7SpkkQ.

563 Richard Eskow, "Is the GOP 'Shock-the-Vote Gang' Planning to Heist California?" *Huffington Post*, February 21, 2006, https://www.huffingtonpost.com/rj-eskow/is-the-gop-shockthevote-g_b_16066.html.

564 Pete Johnson, "Killing Hope: Coverup of the 2004 Election," *freepress.org*, January 3, 2009, https://freepress.org/article/killing-hope-coverup-2004-election.

565 Robert F. Kennedy Jr., "Ohio Election Stolen," *Rolling Stone*, June 15, 2006, reproduced at https://www.commondreams.org/views06/0601-34.htm.

566 Taylor Brodarick, "It's Time to Abolish the Electoral College," *Forbes*, November 12, 2004, https://www.forbes.com/sites/taylorbrodarick/2012/11/04/its-time-to-abolish-the-electoral-college/#12664a5032e0.

567 *King Lincoln Bronzeville Neighborhood Association et al v. J. Kenneth Blackwell et al*, No. 2:2006cv00745 - Document 91 (S.D. Ohio 2009), https://law.justia.com/cases/federal/district-courts/ohio/ohsdce/2:2006cv00745/110360/91/.

568 Brad Friedman, "Lawyer to AG Mukasey: Rove Threatened GOP IT Guru If He Does Not 'Take the Fall' for Election Fraud in Ohio," *Huffington Post*, August 8, 2008, https://www.huffingtonpost.com/brad-friedman/lawyer-to-ag-mukasey-rove_b_115036.html.

569 Bob Fitrakis, "The ghost of rigged elections past: New revelations on the death of Michael Connell," *Free Press*, December 11, 2013, https://freepress.org/article/ghost-rigged-elections-past-new-revelations-death-michael-connell-0.

570 Ben Wofford, "How to Hack an Election in 7 Minutes," *Politico*, August 5, 2016, https://www.politico.com/magazine/story/2016/08/2016-elections-russia-hack-how-to-hack-an-election-in-seven-minutes-214144.

571 "The New Jersey Voting-machine Lawsuit and the AVC Advantage DRE Voting Machine," *Electronic Voting Technology Workshop, Princeton University*, August 2009, https://www.cs.princeton.edu/~appel/papers/appel-evt09.pdf.

572 Leslie Savan, "Last Night's Consolation Prize: Seeing Karl Rove Earn His Nickname 'Turd Blossom,'" *The Nation*, November 5, 2014, https://www.thenation.com/article/archive/last-nights-consolation-prize-seeing-karl-rove-earn-his-nickname-turd-blossom/.

573 Natasha Lennard, "Did Anonymous Stop Rove from stealing the election?" *Salon*, November 20, 2012, https://www.salon.com/2012/11/20/did_anonymous_stop_rove_stealing_the_election/.

574 Michael Hiltzik, "Something's not right about this California water deal," *L.A. Times*, August 18, 2010, http://articles.latimes.com/2010/aug/18/business/la-fi-hiltzik-20100818-1.

575 "Water & Power: A California Heist," *National Geographic*, 2017, https://www.natgeotv.com/za/shows/natgeo/water-power-a-california-heist.
576 Kitty Felde and Viveca Novak, "The Politics of Drought: California Water Interests Prime the Pump in Washington," *OpenSecrets.org*, April 10, 2014, https://www.opensecrets.org/news/2014/04/the-politics-of-drought-california-water-interests-prime-the-pump-in-washington/.
577 Ian James, "Hundreds of water permits expired in national forests," *The Desert Sun*, April 25, 2015, http://www.desertsun.com/story/news/environment/2015/04/25/hundreds-water-permits-expired-national-forests/26361001/.
578 Claire Bernish, "Nestle Pays Only $524 to Extract 27,000,000 Gallons of California Drinking Water," *Antimedia.org*, August 20, 2015, http://theantimedia.org/nestle-pays-only-524-to-exract-27000000-gallons-of-california-drinking-water/.
579 Erica Gies, "Nestle's Thirst for Water Splits Small U.S. Town," *New York Times*, March 19, 2008, http://www.nytimes.com/2008/03/19/business/worldbusiness/19iht-rbognestle.html?mtrref=www.google.com&gwh=6E26C63DE13E2BF9B6A6C389BA3F7F5C&gwt=pay.
580 *Bottled Life* (2012), https://www.bottledlifefilm.com/home.
581 *Tapped* (2009), http://www.tappedmovie.com/.
582 Alexis Bonogofsky, "'Nestle Is Trying to Break Us': A Pennsylvania Town Fights Predatory Water Extraction," *Truth-Out.org*, April 25, 2016, http://www.truth-out.org/news/item/35780-nestle-is-trying-to-break-us-a-pennsylvania-town-fights-predatory-water-extraction.
583 Janice Bobbie, "Water fight—you lose, big business wins," *Economic Democracy Advocates*, February 27, 2017, https://economicdemocracyadvocates.org/2017/02/27/water-fight-you-lose-big-business-wins/.

584 Amber Phillips, "Meet Rick Snyder, the governor at the center of the Flint water crisis," *Washington Post*, January 19, 2016, https://www.washingtonpost.com/news/the-fix/wp/2016/01/19/meet-rick-snyder-the-governor-at-the-center-of-the-flint-water-crisis/?utm_term=.5b05720e5b79.

585 Samuel Osborne, "This is what the poisoned water in Flint, Michigan, looks like," *Independent*, January 16, 2016, http://www.independent.co.uk/news/world/americas/this-is-what-the-poisoned-water-in-flint-michigan-looks-like-a6815661.html.

586 Sanjay Gupta, Ben Tinker, and Tim Hume, "'Our mouths were ajar': Doctor's fight to expose Flint's water crisis," *CNN*, January 22, 2016, http://www.cnn.com/2016/01/21/health/flint-water-mona-hanna-attish/.

587 Allie Gross, "Remember that time Gov. Snyder said he'd drink Flint's water for 30 days straight? That's over already," *Metro Times*, April 25, 2016, https://www.metrotimes.com/news-hits/archives/2016/04/25/remember-that-time-gov-snyder-said-hed-drink-flints-water-for-30-days-thats-over-already.

Chapter 23: Whistleblowing

588 Michael Ray, "WikiLeaks media organization and Web site," *Britannica*, https://www.britannica.com/topic/WikiLeaks.

589 Chase Madar, *The Passion of Bradley Manning* (New York: OR Books, 2012), p. 29.

590 Ibid, p. 29.

591 Ibid, p. 126.

592 Mark Hertsgaard, *Bravehearts: Whistle-Blowing in the Age of Snowden* (New York: Skyhorse, 2016).

593 Denver Nicks, *Private: Bradley Manning, WikiLeaks, and the Biggest Exposure of Official Secrets in American History* (Chicago: Chicago Review Press, 2012).

594 Ashley Parker, "Lawsuit Says Military Is Rife with Sexual Abuse," *New York Times*, February 15, 2011, https://www.nytimes.com/2011/02/16/us/16military.html.
595 *Cioca v. Rumsfeld*, No. 12-1065 (4th Cir. 2013), https://law.justia.com/cases/federal/appellate-courts/ca4/12-1065/12-1065-2013-07-23.html.
596 Jennifer Koons, "Sexual Assault in the Military," *CQ Researcher*, August 9, 2013.
597 Daniel Victor, "'Access Hollywood' Reminds Trump: 'The Tape Is Very Real,'" *New York Times*, November 28, 2017, https://www.nytimes.com/2017/11/28/us/politics/donald-trump-tape.html?mtrref=www.google.com&gwh=1F9278CD15BC87DB0FE700EF2C984799&gwt=pay.
598 Eliza Relman, "The 25 women who have accused Trump of sexual misconduct," *Business Insider*, October 9, 2019, https://www.businessinsider.com/international/women-accused-trump-sexual-misconduct-list-2017-12.
599 Rick Hampson, "Weinstein case fallout: Why now? Why never before?" *USA Today*, November 21, 2017, https://www.usatoday.com/story/news/2017/11/21/weinstein-case-fallout-why-now-why-never-before-rose-spacey-cosby-trump/884385001/.
600 Jacob Pramuk, "Sen. Kirsten Gillibrand calls on Trump to resign over sexual misconduct accusations," *CNBC*, December 11, 2017, https://www.cnbc.com/2017/12/11/sens-gillibrand-merkley-booker-and-sanders-urge-trump-to-resign.html.
601 Martin Boudot, "The Church: Code of Silence (Corrupt Priest Documentary)," *Real Stories*, May 1, 2020, https://www.youtube.com/watch?v=R3Mdd56mx3o.
602 "Is it True that People in the Sea Org Sign a Billion Year Contract," *Scientology.org*, retrieved December 17, 2017, https://www.scientology.org/faq/church-management/is-it-true-that-people-in-the-sea-org-sign-a-billion-year-contract.html.

603 Erin Jensen, "'Leah Remini: Scientology': Former members say parishioners ignored sexual abuse," *USA Today*, August 17, 2017, https://www.usatoday.com/story/life/tv/2017/08/16/leah-remini-ex-scientologists-say-church-members-ignored-abuse/571307001/.
604 Jethro Nededog, "The most shocking allegations of what it's like for children in Scientology, according to Leah Remini's Show," *Insider*, August 16, 2017, http://www.thisisinsider.com/leah-remini-scientology-aftermath-season-2-premiere-recap-child-abuse-2017-8.
605 David Auerbach, "The OPM Breach Is a Catastrophe," *Slate*, June 2015, http://www.slate.com/articles/technology/future_tense/2015/06/opm_hack_it_s_a_catastrophe_here_s_how_the_government_can_stop_the_next.html.
606 Ellen Nakashima, "Hacks of OPM databases compromised 22.1 million people, federal authorities say," *Washington Post*, July 9, 2015, https://www.washingtonpost.com/news/federal-eye/wp/2015/07/09/hack-of-security-clearance-system-affected-21-5-million-people-federal-authorities-say/?utm_term=.7be4ff95dfbc.
607 Cory Bennett, "Calls grow for official to be fired over hack," *The Hill*, July 1, 2015, http://thehill.com/policy/cybersecurity/246612-calls-grow-for-official-to-be-fired-over-hack.
608 Terry Reed and John Cummings, *Compromised: Clinton, Bush and the CIA* (Kew Gardens: Clandestine Publishing, 1995).
609 Barton Gellman, *Dark Mirror: Edward Snowden and the American Surveillance State* (New York: Penguin Press, 2020).

610 Andy Greenberg, "'An NSA Coworker Remembers the Real Edward Snowden: 'A Genius Among Geniuses,'" *Forbes*, December 16, 2013, https://www.forbes.com/sites/andygreenberg/2013/12/16/an-nsa-coworker-remembers-the-real-edward-snowden-a-genius-among-geniuses/?partner=yahootix#40372a1784e4.

611 Conor Friedersdorf, "What James Clapper Doesn't Understand About Edward Snowden," *The Atlantic*, February 24, 2014, https://www.theatlantic.com/politics/archive/2014/02/what-james-clapper-doesnt-understand-about-edward-snowden/284032/.

612 Pete Williams, "US charges NSA leaker Snowden with espionage," *NBC News*, June 21, 2013, https://www.nbcnews.com/news/us-news/us-charges-nsa-leaker-snowden-espionage-flna6C10415586.

613 "Airbus's secret past," *The Economist*, June 12, 2003, http://www.economist.com/node/1842124.

614 Jack Epstein, "Big Surveillance Project for the Amazon Jungle Teeters Over Scandals," *Christian Science Monitor*, January 25, 1996, https://www.csmonitor.com/1996/0125/25071.html/%28page%29/2.

615 James Risen, *Pay Any Price: Greed, Power, and Endless War* (Boston: Houghton Mifflin Harcourt, 2014).

616 James Bamford, "The NSA Is Building the Country's Biggest Spy Center (Watch What You Say)," *Wired*, March 15, 2012, https://www.wired.com/2012/03/ff_nsadatacenter/all/1/.

617 James Ball, "NSA collects millions of text messages daily in 'untargeted' global sweep," *The Guardian*, January 16, 2014, https://www.theguardian.com/world/2014/jan/16/nsa-collects-millions-text-messages-daily-untargeted-global-sweep.

618 Glenn Greenwald and Ewen MacAskill, "NSA Prism program taps in to user data of Apple, Google and others," *The Guardian*, June 7, 2013, https://www.theguardian.com/world/2013/jun/06/us-tech-giants-nsa-data.

619 Katie McDonough, "Glenn Greenwald: Even low-level NSA analysts can spy on Americans," *Salon*, July 28, 2013, https://www.salon.com/2013/07/28/glenn_greenwald_even_low_level_nsa_analysts_can_spy_on_americans/.

620 Russell Brandom, "The CIA has lots of ways to hack your router," *The Verge*, June 15, 2017, https://www.theverge.com/2017/6/15/15812216/cherryblossom-cia-router-hack-surveillance-dlink-linksys-belkin.

621 Glenn Greenwald, *No Place to Hide: Edward Snowden, the NSA, and th U.S. Surveillance State* (New York: Picador, 2015).

622 Luke Harding, *The Snowden Files: The Inside Story of the World's Most Wanted Man* (New York: Vintage, 2014).

623 "Vault 7," *WikiLeaks*, https://wikileaks.org/vault7/.

Chapter 24: Retaliation

624 Kevin Poulson, "NSA Judge: 'I feel like I'm in Alice in Wonderland,'" *Wired*, August 15, 2007, https://www.wired.com/2007/08/nsa-hearing-ope/.

625 Mark Hertsgaard, "Whistle-Blower, Beware," *New York Times*, May 26, 2016, https://www.nytimes.com/2016/05/26/opinion/whistle-blower-beware.html?mtrref=www.google.com&gwh=D599C7B50717493CBB417304BDFF7229&gwt=pay&assetType=opinion.

626 "Text of FBI Director's Remarks on Investigation into Hillary Clinton's Email Use," *New York Times*, July 5, 2016, https://www.nytimes.com/2016/07/06/us/transcript-james-comey-hillary-clinton-emails.html?mtrref=www.google.com&gwh=CC2CF5D3034B8CD288F015EEAFC6BAFC&gwt=pay.

627 "Timeline: The David Petraeus Scandal," *USA Today*, April 23, 2015, https://www.usatoday.com/story/news/nation/2015/04/23/timeline-general-david-petraeus-paula-broadwell-jill-kelley/26245095/.

628 Mary Ilyushina, "Edward Snowden gets permanent residency in Russia—lawyer," *CNN*, October 22, 2020, https://www.cnn.com/2020/10/22/europe/edward-snowden-russia-residency-intl/index.html.

629 Jordan Schatel, "Former Putin Adviser, Russian Media Mogul Found Dead in Washington D.C. Hotel," *Breitbart*, November 6, 2015, http://www.breitbart.com/big-government/2015/11/06/former-putin-adviser-russian-media-mogul-found-dead-washington-dc-hotel/.

630 Jason Leopold, "The U.S. Death of Putin's Media Czar Was Murder, Trump Dossier Author Christopher Steele Tells The FBI," *BuzzFeed News*, March 27, 2018, https://www.buzzfeednews.com/article/jasonleopold/christopher-steele-mikhail-lesin-murder-putin-fbi.

631 Leon Neyfakh, "Report: Russian Propaganda Czar Was Murdered in D.C. Before Planned 2015 Meeting with Feds," *Slate*, July 28, 2017, http://www.slate.com/blogs/the_slatest/2017/07/28/report_mikhail_lesin_russian_rt_founder_was_murdered_in_d_c_in_2015.html.

632 Jamie Grierson, "Litvinenko inquiry: Russia involved in spy's death, Scotland Yard says," *The Guardian*, July 30, 2015, https://www.theguardian.com/world/2015/jul/30/litvinenko-inquiry-russia-involved-spy-death-scotland-yard.

633 Ellen Barry, "Scathing Report Issued on Russian Lawyer's Death," *New York Times*, December 28, 2009, https://www.nytimes.com/2009/12/29/world/europe/29russia.html.

634 Richard Engel and Aggelos Petropoulos, "Lawyer Probing Russian Corruption Says His Balcony Fall Was 'No Accident,'" *NBC News*, July 7, 2017, https://www.nbcnews.com/news/world/lawyer-probing-russian-corruption-says-his-balcony-fall-was-no-n780416.

635 Nicole Sadek, Asraa Mustufa, Dean Starkman, and Hamish Boland-Rudder, "ICIJ's guide to Russian wealth hidden offshore," *International Consortium of Investigative Journalists*, April 20, 2022, https://www.icij.org/investigations/russia-archive/icijs-guide-to-russian-wealth-hidden-offshore/.

636 Claire Duffin, "Billionaire critic of Putin may have been murdered, rules coroner," *The Telegraph*, March 28, 2014, https://www.telegraph.co.uk/news/uknews/10728908/Billionaire-critic-of-Putin-may-have-been-murdered-rules-coroner.html.

637 Sylvia Hui, "Post-mortem shows Russian tycoon died from hanging," *Associated Press*, March 25, 2013, https://web.archive.org/web/20130328102213/http://bigstory.ap.org/article/post-mortem-shows-russian-tycoon-died-hanging.

638 Guilia Carbonaro, "Every Russian Oligarch Who Has Died Since Putin Invaded Ukraine—Full List," *Newsweek*, April 22, 2022, https://www.newsweek.com/every-russian-oligarch-who-has-died-since-putin-invaded-ukraine-full-list-1700022.

639 Katie Davis, "The five clues that suggest four Russian oligarchs were 'all murdered as Putin purges inner circle,'" *The Sun*, April 23, 2022, https://www.thesun.co.uk/news/18351972/clues-putin-allies-murdered-purge/.

640 Jeremy Scahill, *The Assassination Complex: Inside the Government's Secret Drone Warfare Program* (New York: Simon & Schuster, 2016).

641 Christopher J. Fuller, *See It/Shoot It: The Secret History of the CIA's Lethal Drone Program* (New Haven: Yale University Press, 2017).

642 Sonia Kennebeck, "National Bird," *Netflix*, October 13, 2016, https://www.netflixreleases.com/national-bird-2016/.

643 Heather Linebaugh, "Opinion: I worked on the U.S. drone program. The public should know what really goes on," *The Guardian*, December 29, 2013, https://www.theguardian.com/profile/heather-linebaugh.

644 Craig Witlock, "US airstrike that killed American teen in Yemen raises legal, ethical questions," *Washington Post*, October 22, 2011, https://www.washingtonpost.com/world/national-security/us-airstrike-that-killed-american-teen-in-yemen-raises-legal-ethical-questions/2011/10/20/gIQAdvUY7L_story.html.

645 Jeremy Scahill, *The Assassination Complex: Inside the Government's Secret Drone Warfare Program* (New York: Simon & Schuster, 2016).

646 Rob Crilly, "Seven aid workers murdered in Pakistan," *The Telegraph*, January 1, 2013, http://www.telegraph.co.uk/news/worldnews/asia/pakistan/9774124/Seven-aid-workers-murdered-in-Pakistan.html.

647 Bryan Flaherty, "From Kaepernick sitting to Trump's fiery comments: NFL's anthem protests have spurred discussion," *Washington Post*, September 24, 2017, https://www.washingtonpost.com/graphics/2017/sports/colin-kaepernick-national-anthem-protests-and-NFL-activism-in-quotes/?utm_term=.85ca2b71c243.

648 "Trump continues to criticize NFL in tweets," *CBS News*, September 24, 2017, https://www.cbsnews.com/news/trump-criticizes-nfl-sunday-tweets/.

649 Steven Watts, *Mr. Playboy: Hugh Hefner and the American Dream* (New Jersey: Wiley, 2009).

650 John F. Kennedy, "Remarks on Signing Equal Pay Act of 1963," *Presidential Library and Museum*, June 10, 1963, https://www.jfklibrary.org/asset-viewer/archives/JFKPOF/045/JFKPOF-045-001.

651 "Secrets of Playboy," episode 1, *Arts & Entertainment Network*, https://www.aetv.com/shows/secrets-of-playboy.

652 "Secrets of Playboy," episode 3, *Arts & Entertainment Network*, https://www.aetv.com/shows/secrets-of-playboy.
653 Russell Miller, *Bunny: The Real Story of Playboy* (New York: Henry Holt & Co, 1985).
654 "Secrets of Playboy," episode 3, *Arts & Entertainment Network*, https://www.aetv.com/shows/secrets-of-playboy.
655 Holly Madison, *Down the Rabbit Hole: Curious Adventures and Cautionary Tales of a Former Playboy Bunny* (New York: Dey Street Books, 2016).
656 Lanford Beard, "Secrets of Playboy Recounts Horrific Stories About Hugh Hefner's Alleged A-List 'Power Predators,'" *People Magazine*, March 07, 2022, https://people.com/crime/secrets-of-playboy-episode-8-hugh-hefner-predators-ball-bill-cosby-roman-polanski-jim-brown/.
657 Greg Botelho, "Woman alleges Cosby raped her—and other Playboy bunnies," *CNN*, December 5, 2014, https://www.cnn.com/2014/12/05/showbiz/bill-cosby-playboy-accusations/.
658 "Secrets of Playboy," episode 3, *Arts & Entertainment Network*, https://www.aetv.com/shows/secrets-of-playboy.
659 Stefan Tetenbaum, *The Dark Secrets of Playboy*, (Malibu: Around the Way Publishing, 2022).
660 "Secrets of Playboy," episode 1, *Arts & Entertainment Network*, https://www.aetv.com/shows/secrets-of-playboy.
661 Dan Barry, Mike McIntire, and Matthew Rosenberg, "'Our President Wants Us Here': The Mob That Stormed the Capitol," *New York Times*, January 9, 2021, https://web.archive.org/web/20210109232819/https://www.nytimes.com/2021/01/09/us/capitol-rioters.html.

662 Jay Reeves, Lisa Mascaro, and Calvin Woodward, "Capitol assault a more sinister attack than first appeared," *Associated Press*, January 11, 2021, https://apnews.com/article/14c73ee280c256ab4ec193ac0f49ad54.

663 NCTC Cultural History, "The Swearingens Become Prominent Under Lord Fairfax Rule," *U.S. Fish & Wildlife Service*, retrieved November 30, 2017, https://training.fws.gov/history/virtualexhibits/nctcculturalhistory/Timeline1745.html.

664 Paula Wasley, "Back When Everyone Knew How You Voted," *Humanities*, vol. 37, no. 4, 2016, https://www.neh.gov/humanities/2016/fall/feature/back-when-everyone-knew-how-you-voted.

665 Naomi Wolf, "A cure for America's corruptible voting system," *The Guardian*, November 3, 2012, https://www.theguardian.com/commentisfree/2012/nov/03/cure-america-corruptible-voting-system.

666 H. Niles, *Niles' Weekly Register,* vol. 9 (Baltimore: Franklin Press, 1816), p. 214.

667 John Hirst, "Making Voting Secret," *Victorian Electoral Commission*, retrieved August 3, 2017, https://www.vec.vic.gov.au/files/Book-MakingVotingSecret.pdf.

668 Eldon Cobb Evans, *A History of the Australian Ballot System in the United States* (Chicago: University of Chicago Press, 1917).

669 Alexander Keyssar, *The Right to Vote: The Contested History of Democracy in the United States* (New York: Basic Books, 2000).

670 *Bank of the United States v. Deveaux*, 9 U.S. 61 (1809), https://supreme.justia.com/cases/federal/us/9/61/case.html.

671 *Buckley v. Valeo*, 424 U.S. 1 (1976), https://supreme.justia.com/cases/federal/us/424/1/case.html.

672 *Citizens United v. Federal Election Commission*, 558 U.S. 310 (2010), https://supreme.justia.com/cases/federal/us/558/310/.

673 *McCutcheon v. Federal Election Commission*, 572 U.S. 12-536 (2014), https://supreme.justia.com/cases/federal/us/572/12-536/.

Conclusion

674 David Smith, "Trump accuses media of wanting to keep economy shut to hurt his reelection," *The Guardian*, March 25, 2020, https://www.theguardian.com/world/2020/mar/25/trump-accuses-media-of-wanting-to-keep-economy-shut-to-hurt-his-reelection.

675 Lee Fang, "Vaccine Makers Funneled Undisclosed Campaign Cash to Democrats and Republicans in 2020," *The Intercept*, December 14, 2021, https://theintercept.com/2021/12/14/pfizer-moderna-covid-vaccines-2020-dark-money/.

676 Adam Smith, *An Inquiry into the Nature and Causes of the Wealth of Nations* (Chicago: University of Chicago Press, 1976), p. 511.

677 Stephen Quinn and William Roberds, "The Big Problem of Large Bills: The Bank of Amsterdam and the Origins of Central Banking," *Federal Reserve Bank of Atlanta*, August 2005, https://www.frbatlanta.org/-/media/Documents/research/publications/wp/2005/wp0516.pdf?la=en.

www.ingramcontent.com/pod-product-compliance
Lightning Source LLC
Chambersburg PA
CBHW041931260326
41914CB00010B/1253
* 9 7 8 1 7 3 2 1 1 9 7 6 5 *